MW01535150

FOUNTAINS
OF THE
GREAT DEEP

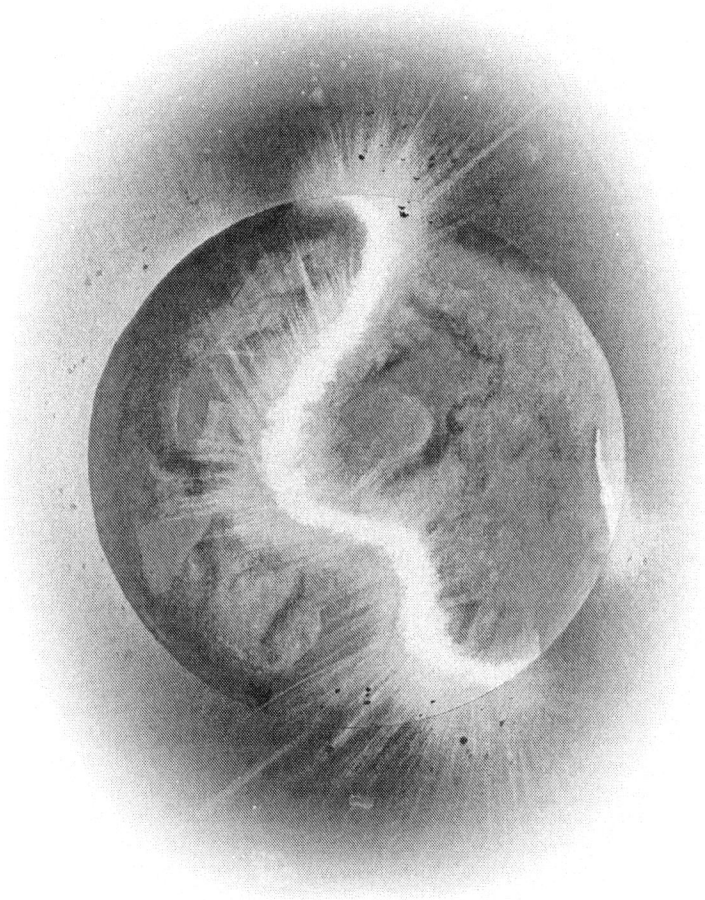

A BRIEF EXPLANATION OF THE
HYDROPLATE THEORY

DIEGO RODRIGUEZ

THE FOUNTAINS OF THE GREAT DEEP
Copyright © 2006
Diego Rodriguez
ISBN 0-9788829-4-6

All Rights Reserved. No portion of this publication may be reproduced, stored in an electronic system, or transmitted in any form by any means, electronic, mechanical, photocopy, recording, or otherwise, without the prior permission of the author. Brief quotations may be used in literary reviews or for the purpose of Bible study.

First Printing - September 2006

Cover Design: Logos Graphics
Cover Illustration © CSC by Steve Daniels

All scripture quotations in this book are from the
King James Version of the Bible unless otherwise identified.

Special thanks to Walt and Peggy Brown for
their help, suggestions, and editorial insight.

Additional copies of this book are available by contacting:

Sound Alive Publishing
PO Box 13008
Fresno, CA 93794
(559) 276-9777

Printed in the U.S.A.

*"By the word of the LORD were the heavens made;
and all the host of them by the breath of his mouth."*
Psalm 33:6

Dedicated to all in the Young Earth Creation Movement,
who strive to edify the body of Christ by increasing our
faith in the authority and infallibility of God's Holy Bible

Table of Contents

Foreword

In this easy-to-read book, Diego Rodriguez has effectively condensed much of the Hydroplate Theory's explanation for the global flood and the supporting evidence. This explanation accounts for twenty four major features seen on earth, as well as the fossil record. Pastor Rodriguez's excellent book has not oversimplified the subject. I highly recommend *The Fountains of the Great Deep* to those of all ages who are curious about this exciting and foundational issue.

Dr. Walt Brown
Author of *In the Beginning*

Introduction

The purpose of this book is to plainly express and concisely explain the *Hydroplate Theory*. The Hydroplate Theory is a scientific theory which attempts to explain the cause, effects, and results of the global flood in the days of Noah. The explanation is based on scientific evidence which harmonizes with Bible scripture. Primarily, the theory seeks to answer the questions: Where did the water (of the global flood) come from, and where did it go? The theory has been proposed by Dr. Walt Brown of the Center for Scientific Creation (CSC), and it is thoroughly and completely explained in his book, *In the Beginning*.

Dr. Brown's book is quite comprehensive and it is full of technical references and mathematical equations to keep even the most learned scientist busy. However, even though the basic portions of his book were written in such a manner that the average high school graduate may understand it, I have encountered many individuals who are interested in the subject, but are intimidated by the sheer size, scope, and content of *In the Beginning*. It seemed good to me to create another book, written at a basic reading level (7th-8th grade as newspapers are written), so that anyone could read and plainly understand the theory. This is the purpose and the reasoning behind this book. This book attempts to plainly describe the content of the Hydroplate Theory by explaining "what happened," without regards to "how or why" these things happened. Those who desire to further inspect and learn the details of the theory should refer to Dr. Brown's book, *In the Beginning* which can be obtained from the Center for Scientific Creation (CSC).

On a personal level, I would like to say that I am thoroughly convinced that the Hydroplate Theory is, at least on a fundamental level, highly accurate. Having looked at the many theories which attempt to explain the global flood, I feel very strongly that the evidence both scientifically and (more importantly) biblically, is overwhelmingly in support of this theory.

Additionally, I believe that the study and explanation of Creation and the origins of life and the world are not only beneficial to Christianity, but essential to the vitality and strength of future generations of the church of the living God. It is my prayer that this book will help to fill the void for those who desire to better comprehend this subject and teach it to others, but have previously felt unable to grasp the levels of technological explanations generally used in such studies.

May the Lord himself bless the reader, and give you a better understanding of God's power and glory.

Pastor Diego Rodriguez

Introduction to Creation Science

In the day and age in which we live, it is very common to hear people say that the Bible does not agree with science. Because of years of false teaching in public education (government schools) and many more years of influence by our popular media institutions, most people have been convinced that the Bible is not scientifically accurate. However, these beliefs are not only incorrect, they are outright lies. Furthermore, these beliefs are rooted in a false foundation and are based upon an improper perspective. The truth is not that the Bible is scientifically inaccurate, but that some so-called "science" is Biblically inaccurate. In short, it is not our responsibility to test the Bible against popular science, but popular science must be tested against the Bible. The foundation that a Christian must have when endeavoring to study science is that the Bible is the final and ultimate authority.

The false foundation and improper perspective referred to earlier is the perspective that scientific discovery is authoritative and scientists are all-knowing and objective. These are the underlying premises that lurk behind every major "evolutionary evidence" or "science contradicts the Bible" statement or article. The problem is that people forget, or have never learned, what science is and what grounds for authority it has, or does not have.

We must remember that science is reported to us in our day by scientists who are fallible men. That means that they are capable of erring. In other words, scientists are not always right, and most of the time, scientists do not all agree with one another. Most public school teachers and media personnel would have you to believe that the stories they present are authoritative and final, and that science (i.e. all scientists) stand behind the story as being true. However, for every theory and discovery, there are always fully credible scientists with matching credentials who do not agree.

You must consider these things whenever you hear of "evolutionary evidence" or "new scientific discoveries" in the news. Many times throughout history, there have been long-held scientific beliefs that were completely overturned when plain and irrefutable new evidence was discovered. This being the case, you must understand that there may be widely held beliefs today, that can and will be overturned as we discover new evidences in the many different branches of science. Therefore, the only unchanging authority we have to place our confidence in is the eternal Word of our God, for He is the Creator of all things. This is why science must line up to the Bible, and the Bible does not need to be changed to fit science!

WHAT IS SCIENCE

In its most basic sense, science is the study of things. Of course, we know and believe that all "things" were created by God. Therefore, true science is the study of God's creation. This study is accomplished through observation, experimentation, and investigation. So, true science will not take away from God's authority, but it will give God glory and increase man's faith in his Creator through the observation and investigation of His creation. True science will give us a better indication of the Creator's power, size, omnipotence, magnitude, strength, omniscience, perfection, providence, and love for man (to name a few). Simple faith allows you to believe things that you don't understand. However, when you have a better insight into how and why you believe and accept the things that you do, it gives you more faith. For example, let us say that a child's parents have told him not to stick his hand on a hot stove. By simple faith, a child will obey the words of his parents. However, when he gets older, he will learn that he cannot put his hand on the stove, because it will burn him. When he learns this, does this increase or decrease his faith in his parents? It increases his faith. It helps him to realize that his parents have more knowledge than him, and can help him in other areas of his life, just as they did with the stove. So it is with science and our Creator.

When we look at God's creation, we can know and believe that God created all things. We can enjoy the beauty and splendor of his magnificent creation. However, when you study God's creation and begin to break down the elements of God's creation through scientific analysis, you will see the perfection and providence of God's perfect designs and the great detail in which all things are created. This will build your faith. You will begin to see how much bigger God is than you previously imagined (He is not just big, He is beyond infinite). When you begin to see the hand of the Creator in all things, you will begin to see his never-ending love and mercy for fallen man. To have an understanding of the detail and exacting perfection of his perfect designs in the earth, sea, in land animals, fishes, birds, whales, horses, the sky, the stars, the moon, the planets, the universe, the human body, the organs of life, our skin, hair, fingers, the structures of society, the marriage bond, the family, and so on… And then to think, all these things were created for the ultimate purpose of God showing his love for us in the plan of redemption through Jesus Christ… this thought is overwhelming.

True science can help bring us to this understanding. However, we must always remember that anything that God creates for good is subject to the corruption of man. Families, churches, sexual reproduction, and even the laws of God are examples of things that have been instituted by God for good use, but have been corrupted and perverted by some people for their own pleasure. Such is the case with science. Science, primarily through the false foundation of evolution, has now been corrupted to the place that society has allowed all manner of vile sin and wickedness to enter in, all in the name of "science." This is not true science, but it is corrupted and perverted science. It is science that is falsely called science. This is the science that the Apostle Paul referred to when he told Timothy to, "…*keep that which is committed to thy trust, avoiding profane and vain babblings, and* oppositions of science falsely so called. *(1 Timothy 6:20)*" True science does not oppose the Bible, but false science does.

The term, "creation science," refers to the area of study in which science is used to give evidences to the creation of all things by God himself, who was the designer and Creator of all things. This book is not intended to give a thorough analysis of creation science, but it may be included in that branch of study, and may be accurately portrayed as a creation science book. For more information about creation science as a whole, please refer to the appendix (page 73) for a list of books and resources on creation science.

The Book of Genesis

The book of Genesis is much more than just the first book of the Bible. The contents of the book serve as the foundation of all doctrines in the Bible. In reality, the rest of the Bible does not make sense, unless we understand the content of the book of Genesis. Truly, the most important events in the history of the world; primarily the life, death, burial, and resurrection of Jesus Christ, do not make any real sense unless you understand the foundations of the book of Genesis.

The word, "Genesis," means "beginning." Therefore, the book of Genesis is the book of the beginning. It is the beginning of the world and the beginning (or origin) of all things, even the beginning of all doctrines. Genesis is not a myth, nor is it an allegorical (symbolic) account, but it is true written history. The stories and elements of Genesis are historic. All of the writings of Genesis are real and true, and the events of Genesis happened in true history. The events of Genesis spanned a history of over 2000 years that began approximately 6000 years ago.

It is impossible to believe in Jesus and the New Testament without believing in the historical accuracy of Genesis. If Jesus and his Apostles' teachings were true, then the historical account of Genesis, which they referred to dozens of times must also be true.

Probably the most important aspect of Genesis, theologically speaking, is the account of the fall of man (Adam). According to the Bible, Adam (the first man), disobeyed God (his Creator). His disobedience is recognized as the first sin, and his sin brought death into the world (Romans 5:12). All men, as descendants of Adam, became subject to sin and death, and it is because of this fact that man needed a Savior. We needed a Savior to deliver us from the bondage of sin and death. We needed Jesus. However, without the historically accurate account of the book of Genesis, we would not understand why there is sin and death in the world. We could not properly understand that we need a Savior. The historically accurate record of the book of Genesis shows us that we do!

GENESIS AND SCIENCE

It is important to note, however, that the book of Genesis is not a science textbook. It is a history book. The reason that it has been subject to scientific scrutiny is because of the initial account of Creation in its first few chapters. We must remember however, that the purpose of Genesis is to explain with historical authority, **what** has occurred, and not what the **scientific details** were in regards to the occurrences.

In the 50 total chapters of Genesis, only the first 11 of them deal with the origins of the world. The remaining chapters deal with the history of the chosen family; that is, the history of Abraham, Isaac, and Jacob. It is easily seen then, that the real emphasis of Genesis is the story of Abraham and his children, and not the scientific details of the account of creation.

However, since Genesis is historically accurate and God is the absolute authority on all things, the account of Genesis chapters 1-11 is not only historically accurate, but it is scientifically accurate as well. True science will show and give evidence that the stories and accounts of Genesis chapters 1-11 harmonize with science.

Therefore, when the Bible says that something happened, we *know* that it happened. The Bible may not tell us about the atmospheric pressure, the coefficient of friction, or the chemical elements involved, but when the Bible speaks, we can know with great certainty that what the Bible said happened, did happen. True science may add insight as to the **scientific details** involved in "what happened," but true science will not take away from *what* the Bible says actually happened. True science will only explain, and may add insight, as to *how* it happened and what the details concerning the event were.

The Hydroplate Theory does just that, and it can therefore be classified as **true science**.

The Hydroplate Theory

The Hydroplate Theory has been proposed by Dr. Walt Brown, who is the founder of the Center for Scientific Creation. Dr. Brown received his Ph.D. from the Massachusetts Institute of Technology (M.I.T.) where he was a National Science Fellow. He has taught college courses in physics, mathematics, and computer science. Dr. Brown is a retired full colonel (Air Force), a West Point graduate, and former army ranger and paratrooper.

In his 21 years of military duty, he served in the following capacities: Director of Benet Research, Development, and Engineering Laboratories in Albany, New York; tenured associate professor at the U.S. Air Force Academy; and Chief of Science and Technology Studies at the Air War College.

Dr. Brown's experiences and training, I believe, made him uniquely capable of proposing and articulating a scientific theory which would explain the global flood of Noah's day. The full theory and its multiple corollaries are completely and technically explained in Dr. Brown's book, *In the Beginning*. Most of the pictures in this book are taken from *In the Beginning*. In fact, on many occasions, when explanations from Dr. Brown's book could not be made any simpler, they were simply restated in this book.

THE PURPOSE OF THE THEORY

The primary purpose of the theory is to explain the global flood in the days of Noah; however, the theory brings insight and additional information to the entire account of the origins of the world in Genesis chapters 1-11.

Starting with the foundation that the Genesis flood is a historically accurate, worldwide event, the theory endeavors to answer the two most obvious questions that such an event brings to the mind:

1. Where did all that water come from?
2. Where did it all go?

A simple answer to these questions could be that God simply created the water. While this is certainly possible (God can do anything he wants), it seems highly improbable. God created all things in 6 days, and on the 7th day he rested. Why did he rest? Did he need to regain his energy after 6 days of creation? Was God tired or exhausted? Of course not. The term *rest* in the context of God and his creation refers to the fact that God rested from his work of creation. In short, God stopped creating things *ex nihilo* (out of nothing). His creation was designed to continue on its own without need for God to specially create anything additional. All living things were created with "its seed in itself" so that living things could continue to produce more living things on their own. The earth, the stars, and the planets, were all perfectly set in motion so as to perfectly continue their rotations without God's need for intervention.

Therefore, since God rested from his creation on the 7th day, it seems theologically inconsistent to conclude that the water of the flood was specially created by God to judge the world. Somehow, the water of the flood was already available for God to use when judgment time came.

The Hydroplate Theory explains where the water came from and where it went with great scientific accuracy and Biblical harmony. Additionally, many other so-called scientific mysteries and phenomena can be explained by the theory. Some of these mysteries and phenomena are:

- The Grand Canyon and other canyons
- Mid-Oceanic Ridge
- Continental Shelves and Slopes
- Ocean Trenches
- Earthquakes
- Magnetic Variations on the Ocean Floor
- Submarine Canyons
- Coal and Oil Formations
- Methane Hydrates
- The Ice Age
- Frozen Mammoths
- Major Mountain Ranges
- Overthrusts
- Volcanoes and Lava

- Geothermal Heat
- Strata and Layered Fossils
- Limestone
- Metamorphic Rock
- Plateaus
- Salt Domes
- Jigsaw Fit of the Continents
- Changing Axis Tilt
- Comets
- Asteroids and Meteoroids

All of these so-called scientific mysteries plainly appear to be the result of a sudden and unrepeatable event. This event was the global flood of Noah's day, whose waters erupted from worldwide, subterranean (under the surface of the earth) chambers, with an energy release exceeding the explosion of ten billion hydrogen bombs!

The Reason for the Flood

Before we discuss the Hydroplate Theory and the scientific evidence for the flood, let us stop to consider one very important truth – the reason for the flood. We know that there was a global flood, and we know that there is plenty of scientific evidence to support it, but what was the reason for the flood? I believe a bit of Bible history is in order here…

In the beginning, God created the heavens and the earth. Within the creation week, on the 6th day, God created man in His own image. We know this man as Adam. The Lord God made Adam a helpmate (a wife) and placed them in the garden, eastward in Eden to dress it and to keep it. This very first couple lived in perfect harmony with God. There was no death, no sorrow, no disease, no thorns, etc. They lived in a paradise on earth. The world they lived in was virtually "heaven on earth." It is uncertain how long they lived in this state, but eventually Adam messed up. Adam committed the very first sin by disobeying God and taking of the fruit of the tree of the knowledge of good and evil. It was at this point that man lost his "paradise," as Adam and Eve were cast out of the garden and the whole universe was cursed as a result of Adam's sin. Women would now bring forth children with great pain and man would labor and toil by the sweat of his brow.

At this point, God told Adam and Eve to be fruitful and fill up the earth (with people). The Bible tells us that Adam begat "sons and daughters," and the generations that came after Adam are plainly declared in the scriptures. Over 1000 years and many generations passed until a man named Noah was born. Noah was a descendant of Seth and the great grandson of Enoch (the man who walked with God and was translated).

According to statistics, there could have easily been anywhere from 100 million to several billion people on the earth in the days of Noah. Unfortunately, the days in which Noah lived were terribly ungodly. We read of the depravity of that generation of people in Genesis chapter 6.

Genesis 6:5 *"And GOD saw that the wickedness of man was great in the earth, and that every imagination of the thoughts of his heart was only evil continually."*

Imagine a day where nobody was righteous. Imagine a time when there were no churches in every city and no missionaries in every country. Imagine a time when literally everybody lived unrestrained without any moral authority. The Bible does not tell us *exactly* what the world was like before the flood, but it does give us this insight:

Genesis 6:11-12 *"The earth also was corrupt before God, and the earth was filled with violence. And God looked upon the earth, and, behold, it was corrupt; for all flesh had corrupted his way upon the earth."*

In some ways, this is very difficult to imagine, but in other ways it is not. It is difficult to think of a time when nobody, except for one man, was righteous before God. Think about it. Imagine a population with millions (possibly billions) of people, and out of all of them, only *one* was righteous. Everybody else had corrupted his way upon the earth.

The corruption is not *completely* difficult to imagine however, since we see this in our own day. The longer you live, the more you discover that people are the same everywhere you go, in every generation that lives. There is nothing new under the sun. All nations, cultures, societies, and generations are subject to the same sins and abominations of their ancestors and neighbors. Just as today, our world is filled with violence, corruption, murder, rape, incest, homosexuality, divorce, hate, cruelty, torture, wars, tyranny, communism, oppression, lies, deception, gambling, drugs, prostitution, devil worship, idolatry, adultery, fornication, anger, perversion, drunkenness, fighting, etc. (it's depressing just to list these things), so was the world in Noah's time. The difference is that everybody, *literally everybody*, except for Noah was partaking in such filthiness. This is very difficult to imagine, but such was the case.

I personally believe that just as we can never fully appreciate and understand the perfect harmony and beauty that existed before Adam's fall, we will probably never be able to fully grasp the horror and terror of the world before the flood.

Because of this great wickedness, God determined to destroy the whole world:

Genesis 6:7 *"And the LORD said, I will destroy man whom I have created from the face of the earth; both man, and beast, and the creeping thing, and the fowls of the air; for it repenteth me that I have made them."*

The Bible makes it very clear that God was not planning on "slapping anybody on the wrist." This was a very serious decision and an even more serious and severe judgment. The Lord determined to completely destroy every living thing that breathed on the earth.

Genesis 6:13 *"And God said unto Noah, The end of all flesh is come before me; for the earth is filled with violence through them; and, behold, I will destroy them with the earth."*
Genesis 6:17 *"And, behold, I, even I, do bring a flood of waters upon the earth, to destroy all flesh, wherein is the breath of life, from under heaven; and every thing that is in the earth shall die."*

This is very important to remember and consider whenever you study about the global flood. Considering that this event was the most severe judgment that God ever passed upon man in the history of the world, one would expect it to be devastating and catastrophic. This is very hard for many people to imagine because they have what I call a "stain on their brain." A "stain on the brain" is a thought or an idea that has been implanted on somebody's mind that they have a hard time getting out, even though it is wrong.

For example, most people think "happy thoughts" whenever they think about Noah and the flood. They think of all the pictures that are seen in children's books, paintings, and art fairs. People think of the cute, smiling animals and the giraffe sticking his long neck out of the porthole in the side of the ark, while Noah stands happily waving his hand as he stands on deck with a beautiful rainbow behind him. Take a look at some of the following pictures as an example (next page):

All of these pictures represent the *stain on the brain* that many people have. Not one, not even *one* of these pictures is Biblically sound, and they definitely do not convey the message of the destruction of the earth by a furious Creator, who was literally fed up with fallen humanity.

Furthermore, the Bible plainly declares that the earth will be destroyed again in a way similar to the way it was destroyed the first time by the flood. Apparently, the world will this time be destroyed by fire and heat:

2 Peter 3:10 *"But the day of the Lord will come as a thief in the night; in the which the heavens shall pass away with a great noise, and the elements shall melt with fervent heat, the earth also and the works that are therein shall be burned up."*

In verses 5-7 of the same chapter of 2 Peter, we read how this coming destruction is likened to the previous destruction of the earth in Noah's day. The Bible teaches here, in this chapter, that one of the signs of the last days will be that people will begin to scoff and mock, and question the certainty of the Lord's return. The Bible then declares that *"this they willingly are ignorant of,"* that just as the whole earth was destroyed by water in a global flood by the word of God, so *"... the heavens and the earth, which are now, by the same word are kept in store, reserved unto fire against the day of judgment and perdition of ungodly men."*

Therefore, anybody who denies the truth of the global flood, and every effort to minimize the catastrophe of the flood is literally fulfilling this prophecy from the book of 2 Peter about the spirit of scoffers in the last days! This is why those children's books with false pictures about the flood, should not be used in a true Christian home. This may sound harsh, but you cannot promote and fill your child with error and expect him to grow up and learn the truth. Fill them with truth, and they will grow up with the truth and love the truth. Fill them with error, and they will grow up with error and turn to confusion.

Ultimately, the most important thing here is to realize that a simple reading of the purpose of the flood in Genesis chapter 6 plainly declares that the judgment which God passed upon "every living thing" was terribly severe and the logical thinker should assume that it was extremely catastrophic!

As you are about to see, the flood may have been more disastrous and cataclysmic than you have ever imagined…

What is a Theory

According to the American Heritage Dictionary, the word "theory" as it applies to the subject at hand has the following definition:

theory *n*. A set of statements or principles devised to explain a group of facts or phenomena, especially one that has been repeatedly tested or is widely accepted and can be used to make predictions about natural phenomena.

In short, a theory is an idea that somebody has come up with to explain things that have been observed. The concept is very simple. For example, let us say that I took a rock and dropped it. I would observe the rock falling to the ground. If I didn't already understand how and why that happened (i.e. gravity), I may come up with an idea, or theory, to explain why things fall when you drop them.

The Hydroplate Theory looks at over 20 major "mysteries" and phenomena (as listed earlier) and explains them all with a simple, yet powerful event: the global flood. This event which was caused by the eruption of water from beneath the surface of the earth triggered a series of events which all explain the current mysteries that we now observe.

HOW TO EVALUATE A THEORY

Since theories are ideas which come from men, there must be a proper way to evaluate them and check their validity, accuracy, and tenability. It is unfortunate that in our day many popular theories are not subject to proper scrutiny or evaluation, but are believed mostly on the grounds of their popularity, or the apparent authority of the scientist(s) who came up with the theory. In other words, some theories, such as evolution, are not properly evaluated for plausibility. They are simply believed by people to be true because many "big-name" scientific authorities support the idea. Also, people often believe in a theory because they *want* it to be true. Other people accept the idea (theory) because it is constantly promoted as being true through the public education process. This is the same way the advertising gurus on Madison Avenue sell us mainstream products and services. Instead of actually educating you as to what the differences are between products in the marketplace and why their product is superior or unique, they simply show us a celebrity or familiar face utilizing the product. Then, they spend millions of dollars repeating the advertising in every medium you can think of, so as to ingrain the image of it on your mind.

This is how evolution and some other theories are propagated. They must be propagated and taught this way because these theories do not stand up to the test of attentive scrutiny. However, this does not have to be the case. There are very simple ways to evaluate theories, and you do not have to be a rocket-scientist, nor a brain surgeon to do it. However, you do have to be open-minded, have common sense, and use logic.

In order to use a theory to explain an unobserved event scientifically, we first have to assume what the conditions were that existed before the event occurred. Assumptions are not made blindly however, nor are they pulled "out of thin air." This is easy to understand. Think about it – if you were not there to witness an event take place, you have to make certain basic assumptions concerning the conditions or surroundings of the event, before the event actually took place. Since you arrived after the event actually took place, you would still be able see the remnants or the leftovers of the event, and you could therefore make some simple assumptions. Once you have determined the basic starting conditions, you then try to determine what the laws of physics dictate would happen.

Police detectives and forensics experts do this all the time. When at the scene of a crime or accident, it is the job of detectives to determine "what happened?" For example, they may not have witnessed a car accident take place, but they know that one did take place because there are two wrecked cars in the middle of the intersection. The detective was not there to witness the accident, so he has to theorize scientifically how this accident took place. He cannot simply say that the red car must have run the red light and hit the blue one because it is a red car, and everybody knows that red cars go fast. He cannot come to a conclusion that suits his fancy simply because he is intellectually lazy or because he has ulterior motives (for example, the driver of the red car is his enemy). He must follow logical steps to come to a logical conclusion based on the evidence and the laws of physics.

First, he would assume certain starting conditions. For example, he may assume, based on what he can see, that the blue car was traveling north and the red car was traveling south. He may assume that the blue car tried to turn onto the perpendicular street at the intersection and the red car hit him. These are logical assumptions that can be made simply by looking at the wreckage and skid marks. However, there is still much to learn. Was the light red when the cars entered the intersection? How fast were they going? Who had the right of way? Is the damage more or less than what should be expected?

Once these answers are scientifically produced using measurements, calculations, and other scientific methods, the explanation (or theory) should then be evaluated based on the following three criteria:

Criterion 1: Process

Process simply means that our assumptions should bring results based on well-established, known processes. For example, if I walked into a field and saw a truck in the middle of a field, I may theorize that this truck arrived there in the field when a squirrel buried a seed in the middle of the field 2000 years ago. Over time, the seed sprouted and turned into this truck. Of course this sounds crazy, but the reason it sounds crazy is because there is no known process that would turn a seed into a truck!

So, a good theory should use explanations that are based upon known processes, or newly discovered processes that can be tested to be true. Theories that call upon "unknown processes" to explain our current observations should be avoided because they are not scientific.

Additionally, explanations should be able to explain many observations. For example, let us say I had observed the scene of the above mentioned car-wreck and took notice of the following:
1. There is a large crack in the radiator of the blue car.
2. The air bag in the red car was deployed.
3. There are two pairs of skid marks traveling north/south.
4. The axle of the red car is broken.
5. The signal lights are not working.

After taking some measurements and making some calculations, my conclusion and description of how the car accident took place should be able to explain all of these various observations. If my description easily explains all of these observations, then we would conclude that my explanation may be correct. However, if my description of the way that the accident happened conflicts with known scientific processes, or if it just does not explain certain observations, then we would conclude that the description is false.

So, if we can easily explain many different observations with an explanation, then our confidence in that explanation should increase. However, if our explanation should have produced certain results, but these results cannot be found, then our confidence in that explanation should decrease. For example, many people have asked the question, "What caused the dinosaurs to die out?" Since this was an unobserved event, that cannot be repeated, we ought to first consider criterion 1. Here is a perfect opportunity to test how to evaluate a theory with this simple criterion...

Some theories that attempt to explain how many dinosaurs "died out," assume that huge climate changes were at the root of their deaths. While it is possible that climate changes may kill dinosaurs, we must also (by criterion 1), consider the other consequences of dramatic climate change. Flowering plants and smaller animals would be even more susceptible to death from dramatic climate changes. Since most of these plants and animals did not "die out," or become extinct with the dinosaurs, then the "climatic change" theories are weakened substantially and should not be accepted. We see then how that theory does not survive simple scientific analysis or scrutiny.

Criterion 2: Parsimony

Parsimony generally means to be extremely frugal or stingy. Some people who would be considered parsimonious may otherwise consider themselves people who are always just trying to get "the most bang for the buck." In other words, they are always trying to get the most production using the least amount of resources as possible.

Science has applied this principle to theories, as well. According to the American Heritage Dictionary, the *law of parsimony* is defined as:

law of parsimony - *Adoption of the simplest assumption in the formulation of a theory or in the interpretation of data.*

In short, all this really means is that it is best to make the fewest assumptions as possible when theorizing. If just a few assumptions can explain many different results, then we should have great confidence in the explanation (and the assumptions). However, if the explanation is so illogical that more and more assumptions are constantly being added to explain why it does not agree with established laws and processes, then we should not have confidence in that explanation.

For example, the theory of evolution is one big assumption. However, before evolution can begin taking place, we must first come to the origin of life question: where did the first form of life come from? Since scientific laws dictate that living organisms cannot arise from non-living matter, believers and supporters in evolution have to continually make more assumptions to explain why their theory contradicts known scientific laws. In the case of the origin of life, the best answer that most evolutionists can come up with is to assume that some "unknown process," caused life to arise from non-living matter. In fact, the whole evolutionary process is a theory which is founded on hundreds of assumptions (that's not parsimony), and even hundreds more *unknown processes*. Is this scientific? No! Evolution will never stand up to the three criteria of effective theory evaluation. It never has and it never will.

Criterion 3: Prediction

Science allows us to make predictions about many things that we do or observe in life. For example, the scientific explanation for gravity would allow us to predict that when you let go of something, it will fall to the ground. However, this is a very simple and oft repeatable event. In fact, it is so common that it is universally accepted and expected. However, where true science really shines is when it is able to predict something that is unusual or would not normally be expected.

A legitimate theory will allow us to make such predictions and demonstrate them to be true when you look in the right places and make the proper measurements and observations. When the predictions are demonstrated and substantiated to be true, then our confidence in the explanation should again be greatly enhanced. *Published predictions are the most important test of any scientific theory.* Anybody can find an unusual piece of evidence and then later claim, "I knew that was going to happen." However, only a studious scientist who has properly examined and evaluated his theory would be confident enough to make multiple published predictions. Few evolutionists make predictions.

Finally, let us note that scientific explanations can never really be certain, absolute, or final. Even the word "prove," is mostly used improperly and is never really justified except possibly in mathematics and in a court of law. Additionally, we must note that science is even less certain when dealing with ancient, unrepeatable events such as the flood and other events in Bible history. Since it is impossible to have *all* of the data and it is impossible to consider *all* of the possibilities and since the events took place *so long ago*, it is possible to overlook certain variables or improperly apply the laws of physics.

Unfortunately, this is the only way to more properly learn about and understand unobservable and unrepeatable events using science. Although we may not be able to explain, nor understand every detail and nuance of a particular event in history, proper scientific evaluation will still allow us to have confidence in certain conclusions.

As stated earlier, let us remember that the Bible is the only authority we have for these ancient events. However, the Bible does not give us scientific detail regarding theses events, it only gives us the historic authority for the events. Therefore, if the Bible declares it, then we can know definitively that it is the truth. Science then, when properly applied, should bring us explanations that support the historical record of the Bible, and also withstand proper scientific evaluations. The explanation then may not necessarily be completely "authoritative," but it should bring us as close as possible to the truth with the currently available data.

In the end, let us remember, that proper scientific process (in this case) will only bring us to a place of great confidence in the theory at hand, but only the Bible can be taken as an absolute, infallible authority.

A Few of the Mysteries

As discussed earlier, there are many mysteries on this earth which current scientific theories do not properly explain. Let us consider some of these so-called mysteries:

Mid Oceanic Ridge. The Mid-Oceanic Ridge is basically a gigantic mountain range. In fact, it is the world's longest mountain range – 46,000 miles long. Since most of it lies on the ocean floor and since it is basically completely under water, many people don't realize that it is there. How did it get there? Why is it primarily on the ocean floor? Why does it intersect itself in a T-shaped junction beneath the Indian Ocean? Why is it made up of rock called **basalt** (pronounced BUH-SALT) and not granite like most other mountain ranges? Why does the portion that passes through the Atlantic Ocean (called the Mid-Atlantic Ridge) seem to almost perfectly divide the Americas from Europe and Africa? If these continents were once connected (as many people believe), then how, why, and when did they break apart?

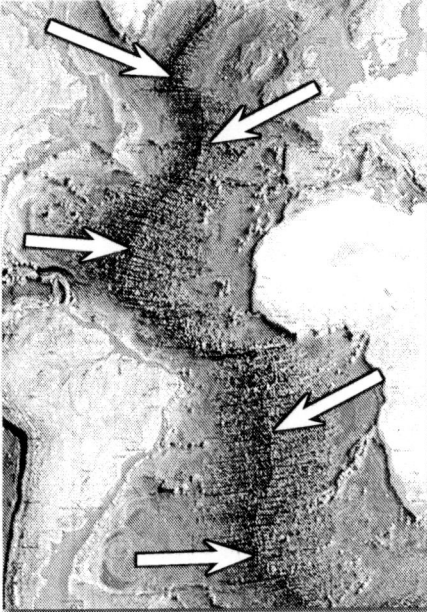

Figure 1 Mid-Atlantic Ridge. This is the portion of the Mid-Oceanic Ridge that passes through the Atlantic Ocean. See the entire ridge on the next page.

Continental Shelves and Slopes. Whenever a continent (like North America) meets the ocean, the continent always extends out a

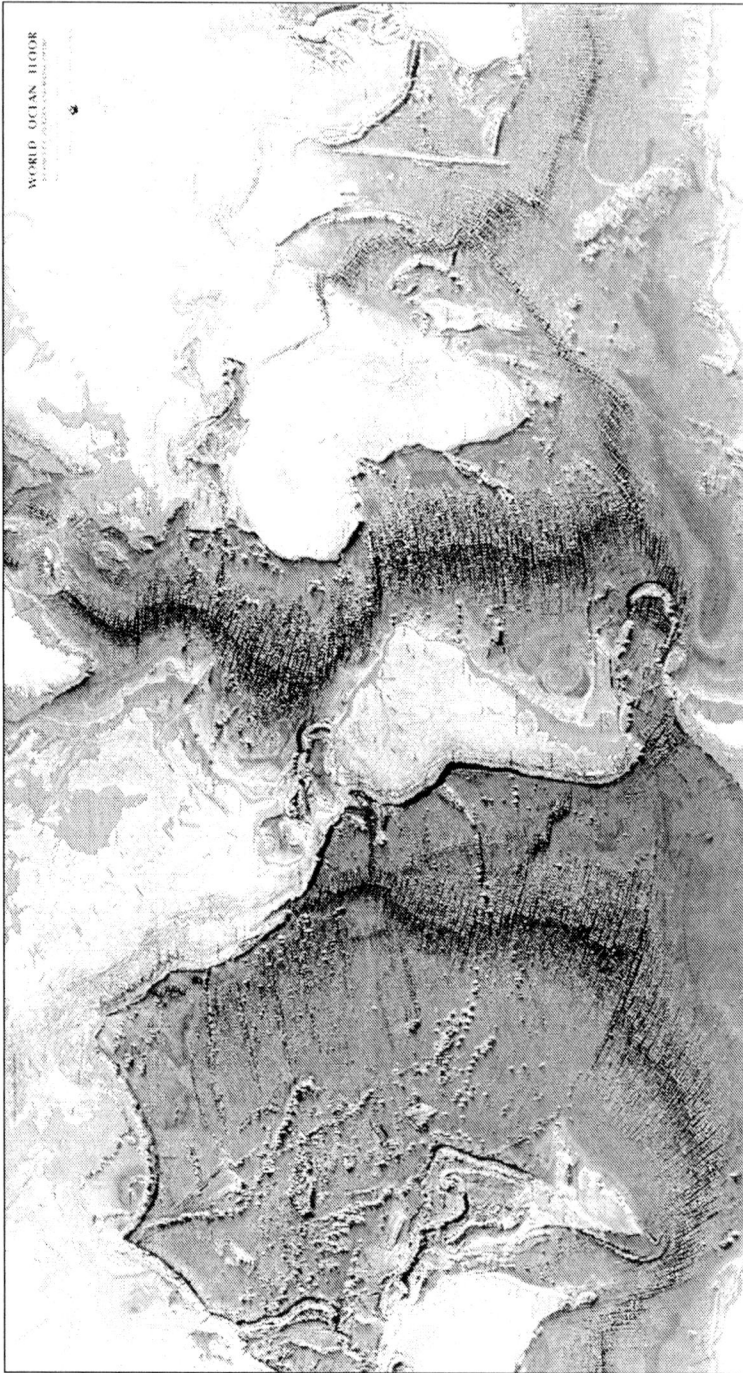

Figure 2 This is a map of the entire world ocean floor. You can see the Mid-Oceanic Ridge as the dark mountain range that runs through the Atlantic Ocean, underneath the tip of Africa, and crosses into a "T" or "Y" in the Indian Ocean (just south of India). Also, it passes through the Pacific Ocean just west of South America and up through North America, where the North American continent actually "runs over" the Ridge.

certain distance into the ocean and then dramatically slopes down to the ocean floor. The edges of all continents all over the earth have this similar slope. Why do all continents slope this way when they reach the ocean?

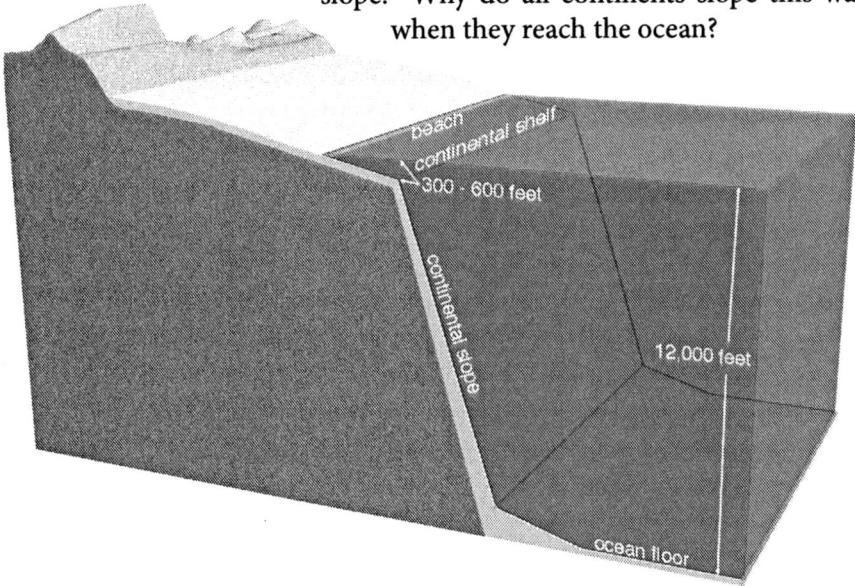

Figure 3 The Continental Margin. Notice how the continent extends out into the ocean before it actually slopes down to the ocean floor. This is the same basic slope that we see on all continental boundaries around the world.

Ocean Trenches. An ocean trench is a long, narrow depression on the ocean floor. Some of them are deeper than the Grand Canyon. Many of them can be seen on the floor of the western Pacific Ocean. When viewed from above, they seem to be folded into the earth along the shape of an arc. When two trenches in the shape of an arc meet, they form a cusp. Why do ocean trenches line the ocean floor with arcs and cusps? What causes them to form in the first place? Why do they seem to be concentrated in

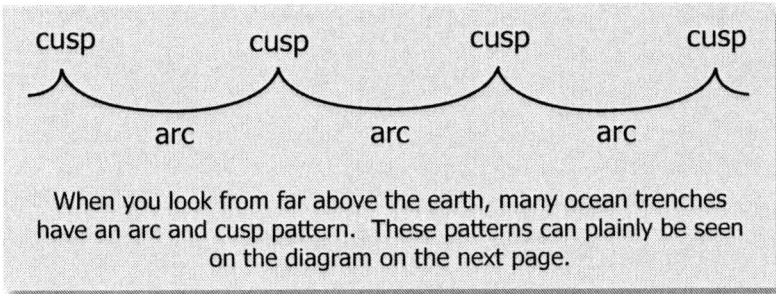

When you look from far above the earth, many ocean trenches have an arc and cusp pattern. These patterns can plainly be seen on the diagram on the next page.

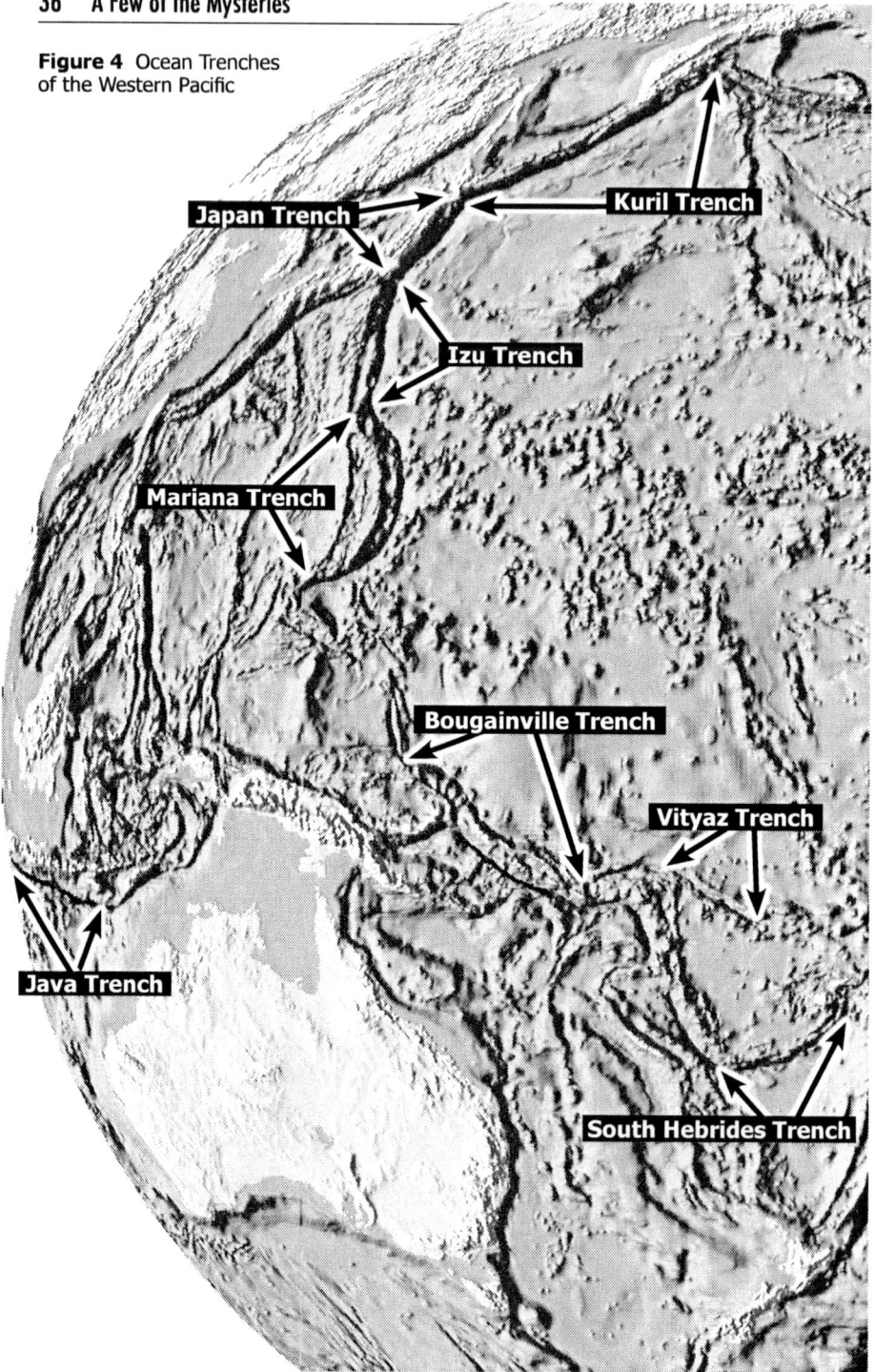

Figure 4 Ocean Trenches of the Western Pacific

the western Pacific floor and not in other parts of the ocean floor?

Submarine Canyons. The ocean floor has hundreds of canyons on its surface. Some of them are larger than the Grand Canyon in both length and depth. One of these canyons is 2,300 miles long – so long that it could almost stretch out across the entire United States. Most of these canyons are in a V-shape and are extensions of major rivers like the Amazon Canyon, the Hudson Canyon, the Congo Canyon, etc. What carved these canyons some 15,000 feet below sea level?

Seamounts and Tablemounts. On the bottom of the Pacific ocean floor, there are thousands of underwater volcanoes, also called *seamounts*. Some are almost as tall as Mount Everest. Strangely enough, the Atlantic Ocean has very few seamounts, but the Pacific Ocean contains approximately 20,000 of them. Why are there so many in the Pacific and so few in the Atlantic? Seamounts with a flat top are called *tablemounts* and they are only about 3,000 - 6,000 feet below sea level. They look like regular volcanoes that had their tops "chopped off." How did these tablemounts get their tops "chopped off?" If the sea level was once much lower, or the ocean floors were higher (or both), and that caused the wave action of the oceans to plane off their tops, then what caused the sea level and ocean floor to come to their present levels?

Coal and Oil Formations. It is generally understood that coal and oil are formed from buried and fossilized plants and organic (living) matter. This matter then decayed in the earth and became the coal and oil that we drill and dig up today. Therefore, in order for coal or oil to form somewhere, there once had to be organic matter there (such as plants, animals, etc.) In fact, many large fossilized trees have been found near both the North and South Poles. In Antarctica, where no trees grow today, some trees found are 24 feet long and 2 feet thick! Nearby are 30 layers of high-grade coal, each 3-4 feet thick. How did so much coal form in a place where no living plants or trees exist today? Well inside the Arctic Circle on the Canadian islands, buried redwood forests with trees 100 feet long have been found. The root structures of these trees show that they grew where they were found. Additionally, much oil is found inside the Arctic Circle. Was it once warm enough for trees to grow inside Antarctica or the Arctic Circle? If it was, then how could so much vegetation (needed to produce all of the coal and oil deposits) grow where it is nighttime 6 months of the year? The vegetation would need daily sunlight. In short, where did all the coal and oil come from?

Frozen Mammoths. Many large elephant-like creatures, called mammoths, and also a few rhinoceroses have been found frozen and buried in Siberia and Alaska. One mammoth still had food in its mouth and digestive tract. This indicates that the animal was frozen very rapidly, almost instantaneously. Had the freezing taken many minutes, we wouldn't find food still in its mouth. What caused this insta-freeze? Furthermore, how did so many large animals find enough food in the Arctic to survive? Since water is frozen up there most of the year, it begs the question, where did these animals find drinking water to survive?

Major Mountain Ranges. If you have ever stood in front of a major mountain range and looked at the mountains intensely, you would have to ask, "how did that happen?" I live in a valley in between two great mountain ranges, and whenever the smog clears to allow us to see these mountains, *it blows me away.* They are just so big! Why do many of them

Figure 5 Folded Mountains. Textbooks and museums often explain the formation of mountains by an unknown force that lifts mountains upward. What is the force? Is that a scientific explanation—or is it a belief? Can you see that an uplifting force could not have created the pattern above? It is evident that the mountains somehow had to be formed by a horizontal force, that squished or compressed the mountains into their shape. Furthermore, since the layers did not crack, they must have been soft like putty when this horizontal compression took place.

look crumpled like an accordion when you fly over them? Some satellite pictures of mountain ranges show that the mountains look like rugs that have pushed up against walls. What force could push a long, thick slab of rock and cause it to buckle and even fold over on itself? Furthermore, even if there was a force that could push such a slab, what would cause the rock to fold or bend and not crack? When you travel through mountains and road cuts on highways (where you can see the mountain layers), why do some mountain layers look like phone books that have been folded over (see Figure 5)? Mountains are hard and they don't bend. If the mountain rock was once like putty so that it could be folded over, what caused it to become like putty, and furthermore, what caused it to become hard again?

Jigsaw Fit of the Continents.
Anybody who looks at the continents on a map or globe can plainly see that they appear to fit together like pieces of a jigsaw puzzle. It is especially evident when you place the bulge of eastern South America just below the bulge of northwestern Africa. For years, people have observed this and wondered how it happened. Sir Edward Bullard proposed the fit that has been included in many textbooks that you see here in Figure 6.

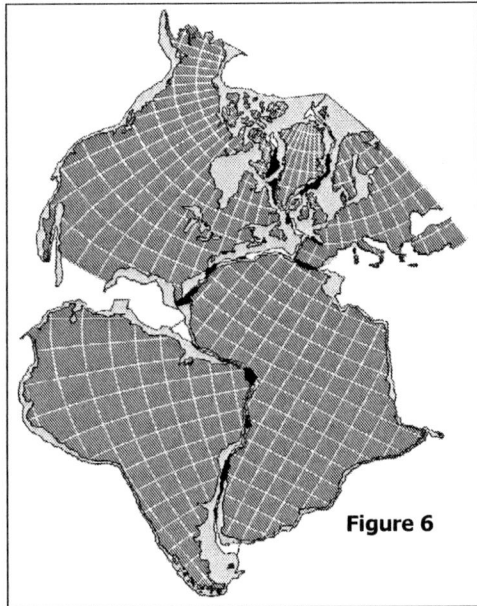

Figure 6

You probably have seen this picture before. Evolutionists claim that these continents have slowly drifted away from each other over millions of years. Well, take a look at the picture. What's wrong with this picture? First of all, notice that Bullard removed Mexico, Central America, and the Caribbean Islands. Also, Europe was rotated counterclockwise and Africa was rotated counterclockwise. Africa's area was shrunk by about 35% and North and South America were both rotated. Of course, the textbooks never inform you of these distortions.

Let's take another look at the continents. First of all, we must remember that the continental boundary (the place where the continent ends) is not the beach, or the ocean shore. The continent usually has a shelf that extends out into the ocean and then drops down into a slope and reaches the ocean floor. This is the true border of the continent. If you were to draw a line around the continents, you would find that they look much different than most people imagine. In fact, the entire Gulf of Mexico virtually rests upon the continental shelf. Furthermore, Asia and North America are actually connecting continents, as are Australia and Asia (see page 34). If you look at the outlines of the continental boundaries, you will see that they almost perfectly align with the base of the Mid-Atlantic ridge. Why is that?

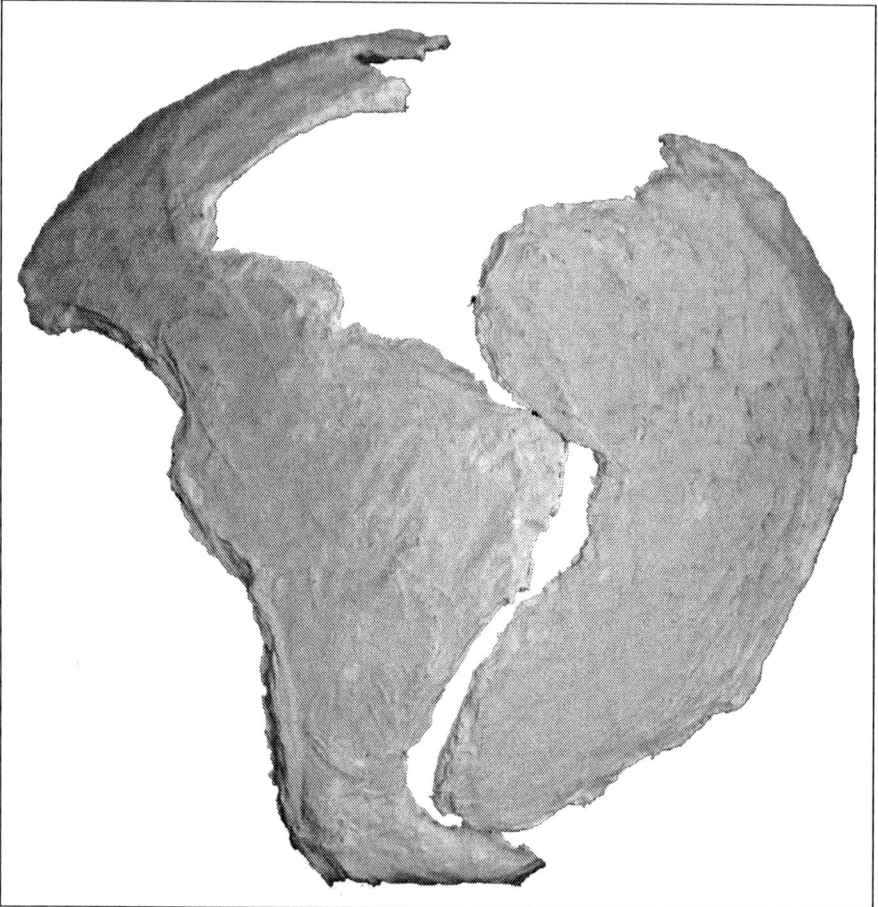

Figure 7 This is a representation of what the actual continents look like. Some of the continents are actually under the ocean water near the continent's edge. As you can see here, the actual continental boundaries do not fit as perfectly as the evolutionists would like you to think. Nevertheless, they still do seem to have been connected at some point in time. The image on the next page shows what the continental plates look like once they are placed on the globe. As you can see, the continental boundaries almost perfectly line up with the base of the Mid-Atlantic Ridge. This begs the question, what happened?

Figure 8

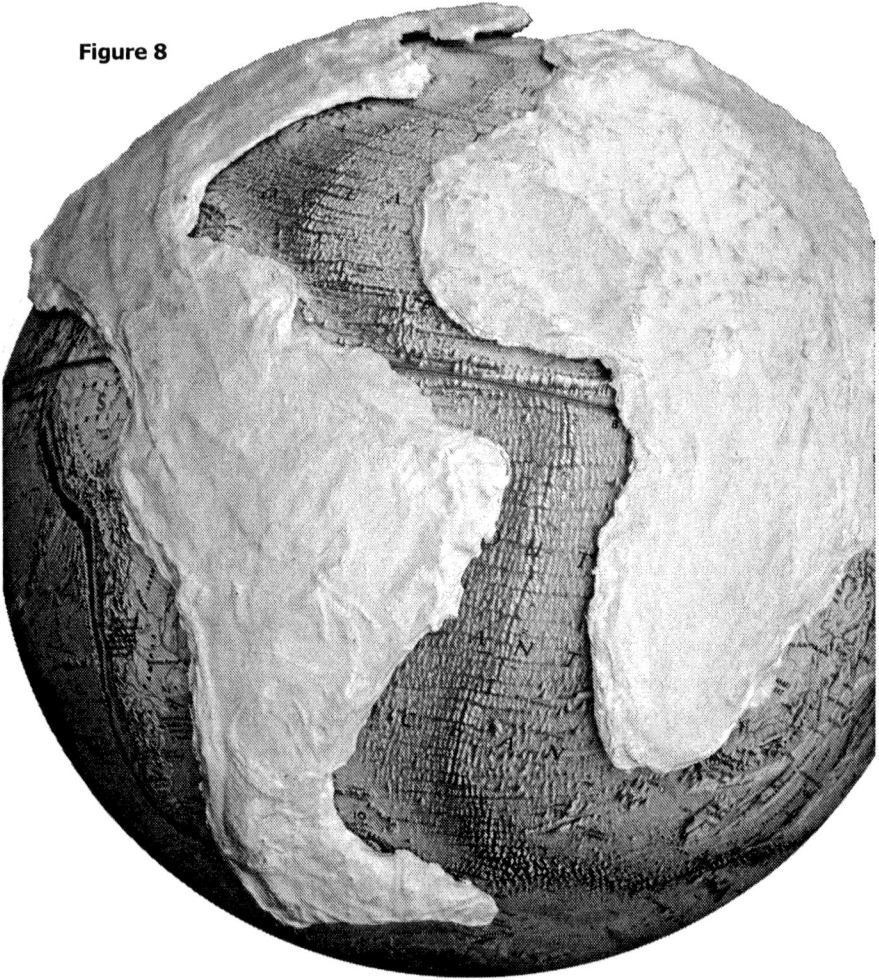

Layered Fossils. Fossils are often claimed to be the greatest evidence for evolution. This is completely insane. Fossils are actually the greatest evidence for Creation and the flood. Think about it. Fossils rarely form today, because living organisms decay much too rapidly. The traditional evolutionary explanation for fossils is that once something dies, it falls to the ground where it is slowly covered by layers of the earth and becomes a fossil over millions of years. Again, the problem here is that no living organism will maintain its form for millions of years. It will completely rot away, be blown away, or be scavenged in a relatively short time. However, there are literally billions of fossils found all over the earth which perfectly preserve the original material. Leaves have been found fossilized which have preserved, with remarkable detail, the veins and structure of the entire leaf. Even soft animals, such as jellyfish have been found fossilized. Fish have been found fossilized in the process of eating other fish, and a marine dinosaur was found fossilized in the process of giving birth. These things evidently could not have taken millions of years. Somehow, all of these organisms were buried very rapidly. Additionally, huge fossil graveyards

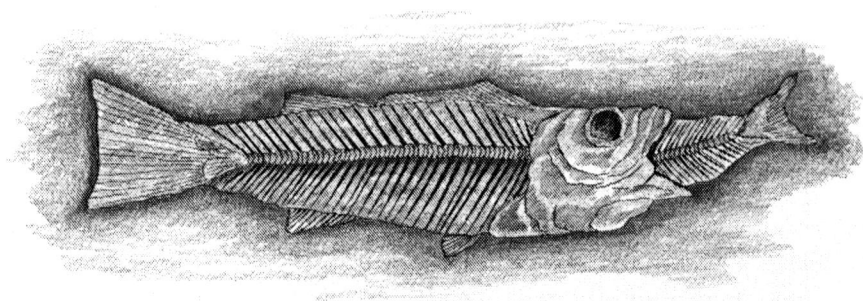

Figure 9 Thousands of fossils like this have been found. How can a fish be captured in the act of swallowing another fish unless it is rapidly buried?

have been found that show stacks of dinosaurs mangled and smashed as if they were all picked up and slammed against the floor. What could have caused this? Furthermore, marine fossils have been found on the tops of every mountain range in the world, and even clam fossils have been found on the top of Mount Everest! How did those marine animals ever get up there? Either the water level was much higher, or the mountains were much lower, or both. In any case, what caused the mountains and oceans to settle at their current altitudes?

The Fountains of the Great Deep

The Bible plainly declares that the entire world was flooded and covered with water in the days of Noah (Genesis chapter 7). One very obvious question that comes to mind when considering this great fact is, where did the water come from? The Bible plainly tells us what the source of this water was in **Genesis 7:11** *"In the six hundredth year of Noah's life, in the second month, the seventeenth day of the month, the same day were all the fountains of the great deep broken up, and the windows of heaven were opened."* Apparently, the major source of water for the flood came from beneath the surface of the earth. The scripture refers to these waters as the "fountains of the great deep." The Hydroplate Theory presumes that the preflood earth was covered with lush vegetation, and that there were seas, lakes, and major rivers, and even small mountains, possibly as high as 5,000 feet.

Beyond that, the Hydroplate Theory is built upon one simple assumption: *there once was a tremendous amount of subterranean water (water underneath the surface of the earth).* Of course, from a biblical perspective, this isn't a complete assumption at all, because as we just read, the Bible plainly declares that this was the state of the preflood world. However, the Hydroplate Theory assumes more specifically that this water amounted to about half of the water that is currently in our oceans and it was located about 10 miles beneath the earth's surface. This water contained a large amount of dissolved salt and carbon dioxide and was contained in large interconnected chambers which altogether comprised a thin spherical shell around the entire earth about ¾ of a mile in thickness. Above this layer of water was a granite crust and beneath it was basaltic rock. To imagine the chambers, picture thousands of very large pools, or pockets of water, which all connect to each other by channels. The only thing that would divide the chambers from each other would be "pillars" or rock structures that connected the granite crust above to the basaltic rock beneath. With this picture, we understand that the earth's granite crust was not *floating* upon these subterranean chambers, but rather was connected to the basaltic rock beneath by the "pillars" in these "chambers."

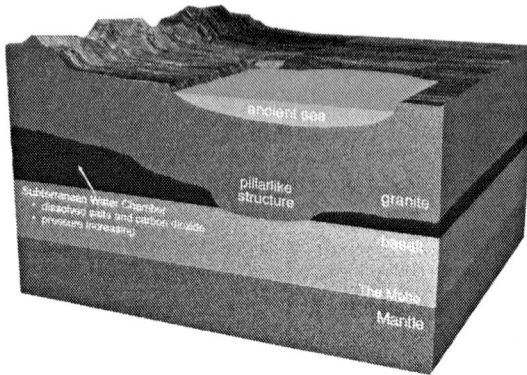

Figure 10 Cross section of the preflood earth. As you can see, the continents would be primarily made of granite, and the subterranean chamber floor would have been made up of basalt. The subterranean chamber was essentially "sandwiched" between granite (above) and basalt (beneath).

Every major mystery discussed earlier and the explanation of the global flood in general, is solved as a result of this simple assumption. So again, we have as our initial condition this one basic assumption: **about 10 miles below the surface of the earth were large reservoirs of salty, subterranean water.**

The chain of events that take place as a result of this *one* assumption will naturally follow known scientific processes, the laws of physics, and will ultimately provide an explanation for the worldwide flood that occurred in the days of Noah, nearly 4400 years ago. The events that took place will be described and related to you as if someone who actually observed these events were telling them to you. These events fall into four phases...

PHASE 1 – THE RUPTURE PHASE

Increasing pressure[1] in the subterranean chamber began to stretch the granite crust of the earth. This is very similar to the way that a balloon stretches or expands when the pressure inside of it increases. As the earth's granite crust continued to stretch, a small crack somewhere in the earth's surface would eventually form. As the pressure continued to increase, this crack in the earth's surface would quickly travel downward and penetrate the subterranean chamber of water. Additionally, the crack would follow the path of least resistance and spread around the earth. The ends of this

---

[1] For an explanation of why the pressure in the subterranean chamber was increasing, please consult *In the Beginning* by Dr. Walt Brown.

Figure 11: Artist's impression of what the jetting fountains may have looked like. For a global perspective, see the artist's impression on the next page.

crack would travel in opposite directions and circle the earth in a few hours. One of these cracks would run into the path left by the other crack and form a "T" or a "Y" on the other side of the earth from where the first rupture began.

As the crack raced around the earth, the 10-mile-thick crust opened like a rip in a tightly stretched cloth. At this point, water would have exploded violently out of the 10-mile-deep "slit" that wrapped around the earth much like the seam on a baseball. The path of this rupture corresponds to today's Mid-Oceanic Ridge. All along this globe-encircling *rupture,* fountains of water blasted supersonically (faster than sound travels) into and far beyond our atmosphere. These fountains gushed out so violently and powerfully that the speed of the flowing water produced repeated sonic booms all over the earth. These supersonic fountains sent literal oceans of water into the atmosphere which then spread out all over the earth and fell back down to earth as rain. This produced torrential rains such as the earth has never experienced – before or after.

Some of the jetting water rose high above the atmosphere where it froze and then fell back down on some parts of the earth as huge chunks of extremely cold, muddy "hail." That hail killed, buried, suffocated, and even froze some animals, including some mammoths. However, the

Frozen
Mammoth
MYSTERY SOLVED

Figure 12: Artist's impression of what the jetting fountains may have looked like from outer space. Notice the asteroids and meteoroids launching out of the earth.

most powerful jetting waters and rock debris were so forcefully launched from the subterranean chamber that they escaped the earth's gravitational pull, and blasted into space to become the solar system's comets, asteroids, and meteoroids.

Comet, Asteroid, and Meteoroid
MYSTERY SOLVED

Some of this launching debris smashed into the near side of the moon creating incredible impact

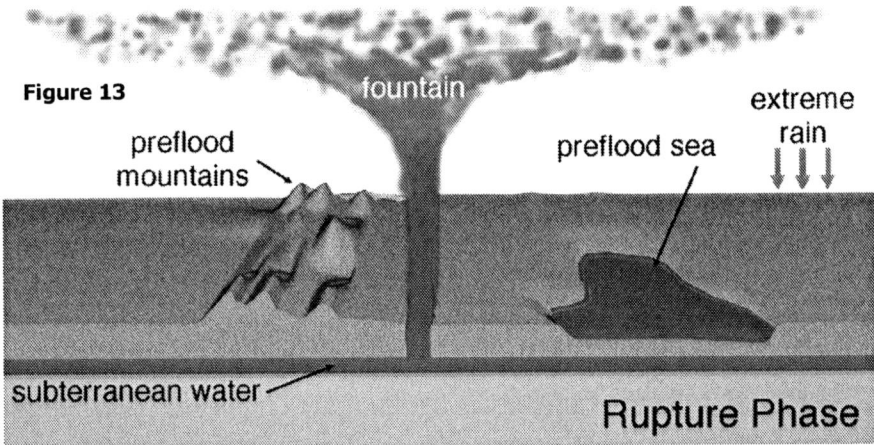

Figure 13

preflood mountains

fountain

preflood sea

extreme rain

subterranean water

Rupture Phase

Prediction – Spacecraft landing on a comet will discover that comets contain elements and sediments from earth, and trace amounts of bacteria and vegetation. The salt concentration will be found to be about twice that of the water in our oceans.

scars, and also producing a great amount of lava which flowed over the moon's surface, smoothing over some craters and producing the dark areas of the moon which gives the moon its "man-in-the-moon" appearance.

Some comets even impacted Mars where they caused temporary saltwater flows. This flowing water is what formed the canyons, gorges, and "erosion" channels that we see on Mars.

Prediction – When the dirt and soil in the "erosion" channels on Mars are analyzed, traces of mixed substances that have been dissolved in water (such as salt from the subterranean chamber) will be found. Soil that is far from erosion channels, will not have these same substances.

PHASE 2 – THE FLOOD PHASE

As the "fountains of the great deep" continued to gush out of the rupture that encircled the earth, they began to rapidly erode both sides of this rupture. The force of the jetting water was so powerful that this rupture eventually eroded to become an 800-mile-wide (on average) chasm that encircled the earth.

Some of the eroded particles (or sediments) from both the basalt chamber floor and also the granite crust were swept up in the gushing waters giving the water a thick, muddy consistency. These sediments then settled out over the earth's surface in a matter of days and instantly buried many plants and animals. This rapid burial process is what eventually formed most of the fossils that we find all over the world.

> **Fossil and Strata (sedimentary layers)**
> **MYSTERY SOLVED**

As the water continued to pour down as torrential rains, the surface of the earth became totally flooded. The rupture continued to gush water, and the rain continued to fall for 40 days. At that time, today's major mountain ranges had not yet formed, so the flooding occurred relatively quickly, since the earth had a smoother surface. Eventually, the rising flood waters covered the jetting water enough to cause the water to stop shooting violently into the atmosphere. At the end of 40 days and 40 nights, the rain ceased. The waters still continued to flow out of the subterranean chamber, however, causing the water level to rise until the 150th day.

The flooding uprooted most of the earth's abundant vegetation. This vegetation was then transported and accumulated in great masses, to various areas of the earth by the currents of the floodwaters. All over the earth, there were large accumulated deposits of preflood vegetation that were buried in the floodwaters.

As the subterranean water raced toward the rupture, it eroded more

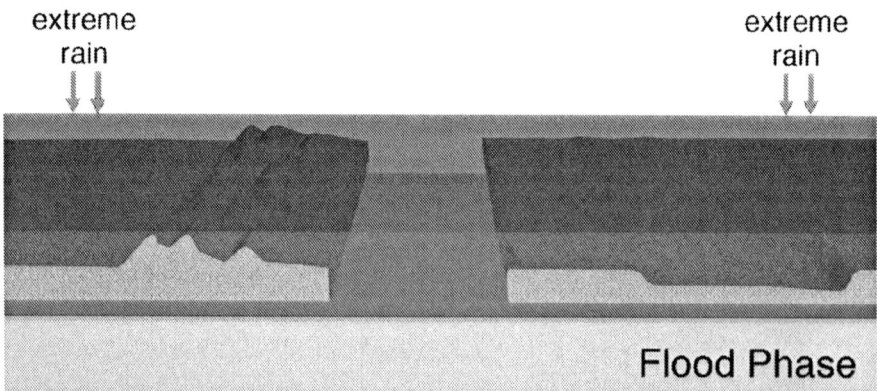

Figure 14 As water began to gush out of the crack in the earth's surface, it eroded the walls and widened the crack until it became more like a chasm. As it widened, the waters continued to gush out and rain back down on the earth as intense torrential rain. This rain continued for 40 days.

of the hydroplate (granite crust) in the direction of the water flow. This caused the hydroplate to sag downward further restricting the flow of the water and increasing erosion even more. This sagging and eroding caused the hydroplate to be beveled (sloped or angled), and eventually formed the familiar shape of the continental shelves and slopes we see all over the world.

Continental Shelves and Slopes
MYSTERY SOLVED

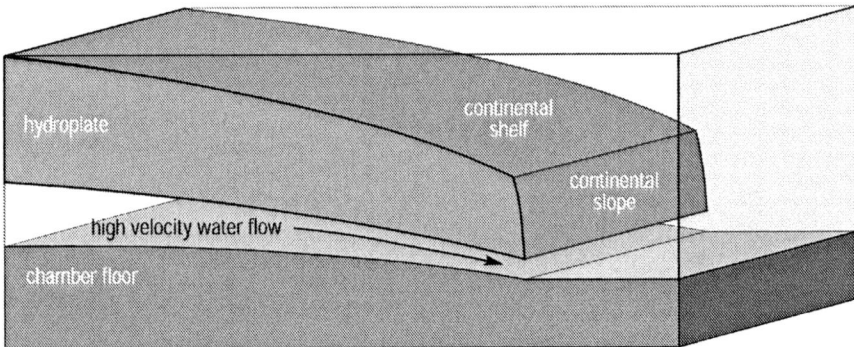

Figure 15 As the subterranean water raced horizontally towards the rupture, the overlying hydroplate eroded more rapidly in the direction of the water flow. This caused the hydroplate to sag in that direction which ultimately formed the beveled and sloped continental shelves and slopes we have today.

PHASE 3 – THE CONTINENTAL DRIFT PHASE

If you place weight on a rock, it will slightly compress the rock like a spring. Rock that is deep within the earth has a considerable amount of weight upon it, so it can be compared to a tightly compressed spring, in a sense. This compression increases with depth all the way down to the earth's core.

As water continued to flow out from the subterranean chamber, the walls of the crack increasingly eroded. Eventually, the width of this chasm was so great, that the compressed rock on the chamber floor sprung upward. It did so because the weight that once pressed down upon it had now been removed, and the chamber floor was now exposed all along the rupture.

This upward rising first began in the area we now recognize as the Atlantic Ocean, and it gave birth to what we now call the Mid-Atlantic Ridge. As the Mid-Atlantic Ridge rose, it created slopes on both sides of

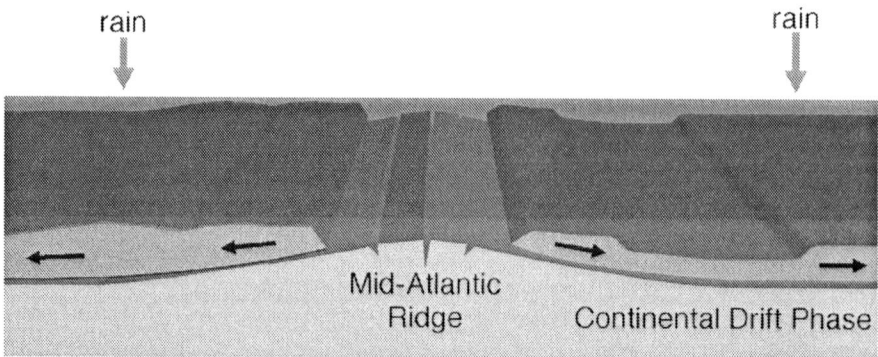

Figure 16 During the continental drift phase, the weight of the continents pressing down on the chamber floor, combined with the widening of the "chasm" caused the ocean floor to begin to rise–forming the Mid-Atlantic Ridge. For another example of how and why this occurs, see Figure 16 on the next page.

itself. These slopes lifted the granite plates, which we will call *hydroplates[1]*, causing them to slide down the slopes. As the hydroplates slid downhill, they further widened the chasm, which removed more weight from the chamber floor, causing the Mid-Atlantic Ridge to rise faster. This process continued until the entire Atlantic floor rose almost 10 miles.

When the Mid-Atlantic Ridge began to rise out of the ocean floor, it also lifted portions of the chamber floor which were north and south of it. This caused those portions to become unstable and also spring upward, starting a chain-reaction process which continued all along the rupture path forming what we know today as the Mid-Oceanic Ridge (see Figure 16 on the next page). Today, the Mid-Oceanic Ridge is the longest mountain range on the earth, and it completely wraps around the earth, as previously stated, like a seam on a baseball. The part of the Mid-Oceanic Ridge that passes through the Atlantic Ocean is called the Mid-Atlantic Ridge, although it is part of the entire Mid-Oceanic Ridge.

> **Mid-Oceanic Ridge**
> **MYSTERY SOLVED**

The hydroplates, lubricated with the subterranean water that was still escaping from beneath them, continued to slide away from the rising Mid-Oceanic Ridge. In time, these hydroplates ran into two different types of resistances. The first resistance was brought about when the lubricating water beneath the hydroplate was depleted. The second resistance was caused when a hydroplate crashed into something. As each gigantic hydroplate decelerated by either cause, it experienced a great *compression*

---

[1] Now you know what a **hydroplate** is! The term "hydro" means water. So, a hydroplate, is a plate that rests upon water. Dr. Brown coined this term.

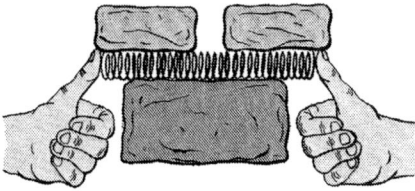

The weight of these rocks will keep the spring in its horizontal position.

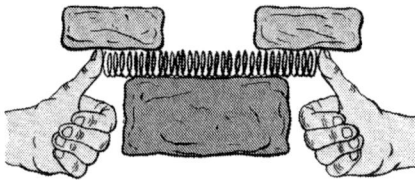

Rupture has formed. Gushing water is not shown in this image.

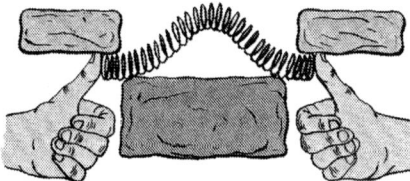

As the distance between the rocks increases, the spring stays horizontal.

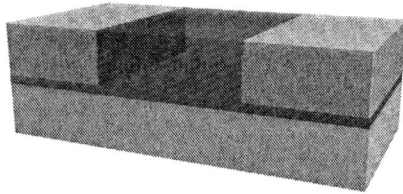

The "chasm" increases in width because of the eroding power of the gushing water. Most of the Earth's sediments were created by the walls of this gap being eroded by the "fountains of the great deep."

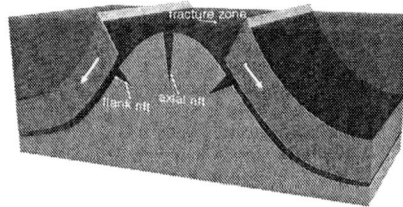

Once the distance between the rocks reaches a certain width, the spring will suddenly buckle upward. Now imagine thousands of springs lined up parallel to each other with rocks on top of them also. If the coils were all attached to each other, then as soon as one buckles upward, because of the growing gap between the rocks, it will start a chain reaction causing all of the other springs to buckle upward as well. This process would continue down the entire range of springs in a ripple-like fashion.

Figure 17

Continental Drift phase begins. Once the "chasm" grew to about 800 miles wide, the Mid-Atlantic Ridge began to form as the subterranean chamber floor buckled upward. This caused the massive hydroplates to begin sliding downhill as they rode on a layer of lubricating water. As the weight of these hydroplates is removed from what will become the Atlantic floor, this exposed floor quickly rises almost 10 miles, forming the long Mid-Atlantic Ridge that we know today. Since the Ridge is forming and stretching upward, it is easy to see how and why the axial rifts and flank rifts formed.

event – buckling, crushing, and thickening each plate. Imagine taking a large brick of soft clay. Now, imagine that you have placed the clay brick on a table which is lubricated. Finally, imagine yourself pushing the clay brick along the table and into the wall. When the brick hits the wall, it will *compress*, and it will buckle, crush, and thicken. Essentially, this is what happened to the hydroplates. The compression event easily and continually crushed and thickened each hydroplate for many minutes. This process quickly squeezed up mountains all over the earth.

> **Mountain Formation**
> **MYSTERY SOLVED**

Naturally, the mountains would form parallel to the oceanic ridges from which they slid. If we were to imagine the clay brick mentioned earlier being pushed by our open hand up against the wall, it would buckle upward and form a "mini mountain range," so to speak, which would be parallel to our hand pushing it. This is basically what happened with the sliding hydroplates, and again it explains why the major mountain ranges are parallel to the portions of the Mid-Oceanic Ridge from which they slid. In fact, if you look at the ocean floor map on page 34, you will see that the major mountain chains generally run parallel to the Mid-Oceanic Ridge.

As the mountains buckled up, some of the remaining water underneath the hydroplate would have flowed under the mountains to fill in large gaps caused by the compression event. Some pooled water should still remain underneath cracked and contorted layers of rock.

> **Prediction** – *Beneath major mountains are large volumes of pooled saltwater.*

> **Prediction** – *Salty water frequently fills cracks in granite 5-10 miles below the earth's surface, where surface water should not be able to penetrate.*

As the hydroplates began to decelerate rapidly due to resistances, the bottom of the hydroplates skidded along the surface of the subterranean chamber. This skidding produced intense friction, which generated enough heat to melt rock, which in turn, produced a tremendous amount of magma (molten rock). The crushing during the compression event produced similar results, as broken and extremely compressed rocks slid past each other. The amount of friction and heat was intense. In some places, the extreme temperatures and pressures formed metamorphic rocks, like diamonds and marble. Any magma that was under great pressure would "squirt" through cracks in the crushed rock. Sometimes, magma would escape to the earth's surface producing volcanic activity and "floods" of lava. This was the beginning of the earth's volcanic activity. Still, other pools of magma collected together under the earth's surface

Origin of Volcanic Activity
MYSTERY SOLVED

in pockets, which we now call magma chambers. When the magma escapes from these chambers, volcanoes erupt. The heat which remains today from this activity is called geothermal heat.

As the Mid-Atlantic Ridge and the Atlantic floor rose, mass had to shift within the earth and move itself toward the Atlantic. This mass shifted from the opposite side of the earth. Essentially, mass moved within the earth, shifting slightly from the Pacific side of the earth toward the Atlantic.

Formation of Ocean Trenches
MYSTERY SOLVED

This caused portions of the Pacific Ocean to cave in or sink in on itself. This "caving in" process is called *subsidence*. It was this subsidence that formed huge ocean trenches in the Pacific Ocean.

Since the formation of ocean trenches took place after most of the vegetation and land animals were buried, we would expect to find fossils of these buried land animals even in deep ocean trenches.

> ***Prediction*** – *Fossils of land animals, not just shallow-water sea creatures, will be found in and near ocean trenches*

Additionally, during this phase, the buried layers of vegetation that accumulated in the previous phase would have been rapidly compressed and heated, which are exactly the conditions required to form coal and oil. The large deposits of coal and oil that we have today came from these events.

Coal and Oil Formation
MYSTERY SOLVED

PHASE 4 – THE RECOVERY PHASE
(Where did all the water go?)

When the compression event took place on each individual hydroplate, the hydroplate would compress, thicken, form mountains, and rise up out of the water. As this happened, the water receded back into the huge basins that opened up as a result of the hydroplates compressing and thickening.

At the same time that this was happening, the outward gushing waters from the subterranean chamber were "choked off" as the hydroplates settled onto the chamber floor. Since the hydroplates were no longer sinking, water was no longer forced up onto the earth. Once this source of water was shut off, the new and deep basins that formed between continents basically became huge reservoirs into which the flood waters returned. This happened all over the earth, giving birth to our present oceans.

As the flood waters drained down the steep continental shelves to the lowered sea level, it eroded deep gorges, which today we call *submarine canyons*. These canyons are especially dominant downstream of drainage channels which are now major rivers.

**Submarine
Canyon Formation**

MYSTERY SOLVED

For years after the flood, sea levels were considerably lower than today, because the thickened continents had not yet settled into the earth's mantle. However, over time (possibly a few hundred years), as the hydroplates did "settle in" to the chamber floor, the ocean floors had to rise to compensate for the settling hydroplates. This is similar to what would happen if you were to cover a waterbed on one side with a granite slab and the other side with a blanket. The granite slab would sink causing the blanket to rise. This process also took place on the earth. Immediately after the flood, there were large "land bridges" that connected the major continents. North America was connected to Asia through what is now the Bering Sea, and Australia was connected to Asia. In fact, you can see on a map of the ocean floor (page 34) how all major continents would connect if the sea level was lowered just 300 feet.

As sea levels rose over the next few centuries, the continents were eventually *divided by water*. Since the rising water level ultimately cut off the land bridges between continents, some groups of people and species of animals became isolated from the "rest of the world," and continued their existence in their new "native" lands. It was most likely during this time that Eber named his son Peleg (Genesis 10:25), because Peleg means a *division by water*.

Over time, the new mountain ranges that had formed also began to exert more pressure on the chamber floor that was beneath them. As the mountains slowly settled into the earth, parts of the earth that were adjacent to the mountains rose, forming plateaus. This is similar to the manner in which the ocean floors rose as the hydroplates settled into the chamber floor (or the upper mantle), and it is also similar to the way a waterbed rises on one side when you lay down on the other side and

**Plateau
Formation**

MYSTERY SOLVED

sink into it. This also explains why plateaus are always adjacent to major mountain ranges. For example, the Tibetan Plateau (the largest and tallest in the world), is next to the most massive and highest mountain range in the world – the Himalayas. Other examples of this are the Colorado Plateau which is next to the Rocky Mountains, and the Columbia Plateau which is next to the Cascade Mountains.

The sudden formation of the earth's major mountain ranges disrupted the earth's spinning balance. Since the earth is spinning, large mountains sticking out on the surface of the earth would throw off the earth's balance. This is somewhat like the vibrations caused when an offset weight is attached to a small motor (note–these types of vibrating motors are currently used in cell phones and pagers). Ultimately, this caused the earth to actually roll about 35-45 degrees. The preflood North Pole moved to what is now central Asia. This is why coal is found at today's South Pole, and it is also why so many researchers have found lush vegetation, vast dinosaur remains, and frozen mammoths inside the Arctic Circle. These locations were at temperate latitudes before the flood.

As the waters that once covered the earth drained into the ocean basins, they naturally would have left every bowl-shaped depression in the earth filled to the brim with water. The compression event would have created many postflood basins that would act like "bowls" and hold water. These "bowls" became large postflood lakes of various sizes and shapes. Over time, some of these postflood lakes would lose much of their water through evaporation and seepage. In such cases, the lakes would shrink over the centuries. A well-known example of this was former Lake Bonneville, part of which is now the Great Salt Lake.

However, other postflood lakes would gain water over the centuries from rain and from drainage from higher terrain. This would force the water in these lakes to flow over the rim of these lakes at the lowest point on their rims. This would then erode that part of the rim where the water flowed, causing even more water to flow over it, which in turn would increase the erosion even more. This erosion would begin to accelerate dramatically and catastrophically. Eventually, the entire lake would dump out of a deep slit formed by this erosion. Today, we call those slits *canyons*. Waters from these lakes would then spill into the next lower basin (or lake), causing it to overflow its rim, and repeat the same erosion process. It was similar to falling dominoes. The most famous canyon of all, the Grand Canyon in northern Arizona, formed primarily by the dumping of a postflood lake called **Grand Lake**. This lake sat upon the southeast quarter of Utah, parts of northeastern Arizona, and small parts of Colorado and

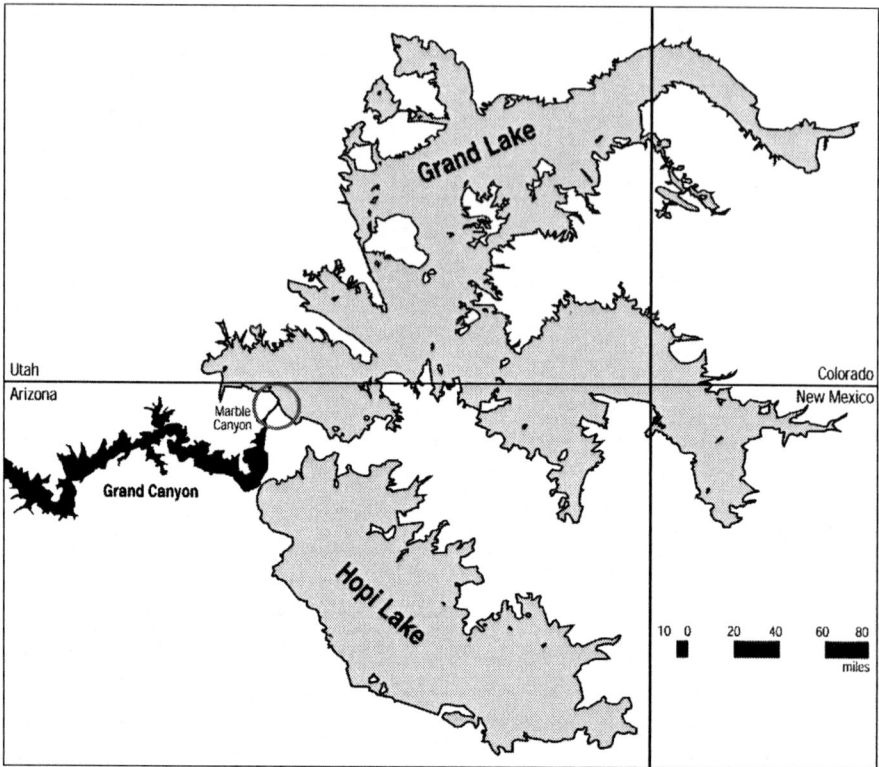

Figure 18: Grand Lake (named by Dr. Walt Brown) and Hopi Lake. These lakes no longer exist. However, it was the water from these lakes that poured out over northern Arizona in a matter of weeks and carved out the Grand Canyon.

New Mexico. Grand Lake, which sat at an elevation of about 5,700 feet above sea level, spilled over its rim and quickly eroded its natural dam 22 miles southwest of what is now the town of Page, Arizona. In doing so, it eroded the western boundary of former Hopi Lake, releasing the waters that occupied what is presently the valley of the Little Colorado River. In only a few weeks, probably more water was released over northern Arizona than is in all five Great Lakes combined. It was the massive release of this water in a short period of time that carved out what we see today as the Grand Canyon. The next time you see a picture of the Grand Canyon, or visit it yourself, you can stop and consider that what you are looking at is literally the leftovers, or the aftermath, of the wrath of God.

Since thousands of postflood lakes would have been formed, these processes took place many times all over the globe and formed many canyons that we can see today. Additionally, by looking at other locations on the globe, we can easily conclude the following:

1. The Mediterranean "lake" dumped into the lowered Atlantic Ocean and carved out a canyon at the Strait of Gibraltar.
2. A giant lake which filled the entire Central Valley in California (where I live), carved a canyon (now largely filled with sediments) under what is now the Golden Gate Bridge in San Francisco.
3. The Mediterranean Sea or the Black Sea carved out the Bosporus and the Dardanelles in Turkey.

> **Prediction** – *The rock under the Gibraltar, the Bosporus and Dardanelles, and the Golden Gate bridge will be found to be eroded into V-shaped notches which would have formed when waters spilled over the rim of the original postflood lakes. (This prediction concerning the Bosporus and Dardanelles, which was originally published in the 1995 edition of In the Beginning, was confirmed in 1998).*

Unquestionably, the global flood was a catastrophe of unimaginable magnitude and force. The words that we use today to try to describe an event like this, such as: cataclysm, catastrophe, or disaster – do not seem to relay the true magnitude of this earth-shaking historical event.

The Bible and the Hydroplate Theory

Although the Hydroplate Theory is a complete scientific theory, it harmonizes remarkably with the text of the Word of God. Again, we must remember that true science will always harmonize and confirm the Word of God. The Hydroplate Theory does just that. Additionally, if we were to accept the events and explanations of the Hydroplate Theory as generally being true, then it appears it may help to give us some additional insight into certain scriptures.

Here, we will discuss the harmony of the Hydroplate Theory with the Bible, and discuss some possible insights that we may be able to glean from the theory's various explanations.

THERE ONCE WAS A GREAT AMOUNT OF WATER UNDER THE SURFACE OF THE EARTH

Many scriptures mention water underneath the earth. It is generally accepted and understood that this is the way the world was first created. The following scriptures are easily understood and described by the theory:

Psalm 24:2 *"...for he founded it [the earth] upon the seas and established it upon the waters."* NIV

This scripture evidently speaks of water underneath the surface of the earth, for it declares that God established the earth "upon" the waters. Obviously, this would mean that the earth's surface somehow rested upon water. The Hydroplate Theory explains how this could be.

Psalm 33:7 *"He gathers the waters of the sea together as a heap; He lays up the deeps in storehouses."* NAS

This scripture seems to convey an interesting meaning. Here, the Bible explicitly declares that God "lays up" or puts away the deep ocean waters in storehouses. A storehouse is normally understood to be a large repository that keeps, stores, or preserves something for later use. In this case, it appears as though God originally made the earth with subterranean water (that the Bible poetically describes as being the deep water that was kept in storehouses), which God kept available for his later use. Of course, we understand that this water was used on the 17th day of the 2nd month, 1656 years after creation, when the fountains of the great deep broke open!

Psalm 104:3 *"He lays the beams of His upper chambers in the waters..."*

Psalm 136:6 *"To Him that stretched out the earth above the waters..."* NAS

These scriptures are interesting because they seem to denote the fact that the earth rests upon a layer of water and is supported by "beams." What were these beams, and how did they support the earth?

According to the Hydroplate Theory, the preflood earth had a relatively smooth surface. However, we know that there were some hills, because the Bible mentions them in Genesis 7:19. These hills would have been rather small compared to today's mountains. We can easily understand then, that the thickness of the crust of the preflood earth was not perfectly uniform. In other words, there were some parts of the earth's crust that were thicker than other parts. In certain areas, particularly areas where hills were, the crust would "sag" down into the chamber. Rock that is under the pressure of 5 miles of rock or more sitting on top of it, will flow like putty. Since the earth's crust was about 10 miles thick, these "sagging" areas of the crust would easily droop down and press against the floor of the subterranean chamber.

This rock that squeezed down and reached the subterranean chamber floor would have looked something like a thick icicle or stalactite. Thousands of these "pillars" would have existed in the subterranean chamber. These "pillars" would correspond to the "beams" that we read about in Psalm 104:3. The Bible speaks in other places of the earth being formed or founded upon pillars, and it is highly likely that these subterranean pillars are those pillars described:

Psalm 75:3 *"The earth and all its inhabitants are dissolved; I [God] set up its [the earth's] pillars firmly."* NKJV

Here, God testifies to the fact that He is the one who set up the pillars of the earth. Apparently, these were the pillars in the subterranean chamber upon which the earth rested. In fact, it is evident that the earth had to rest upon something like pillars, or else the earth's surface would continually totter and wobble. It appears as though God established pillars, or beams, in the subterranean chamber, upon which the earth's crust rested to keep the earth's surface from rolling and tottering:

Psalm 104:5 *"He established the earth upon its foundations, <u>So that it will not totter</u> forever and ever."* NAS

In Job 38, God shows Job his sovereignty by describing his Creation and asking Job to explain certain aspects of Creation that only God could know about. In verses 4-6 God asked Job:

Job 38:4-6 *"Where wast thou when I <u>laid the foundations of the earth</u>? declare, if thou hast understanding...Whereupon are the foundations thereof fastened? or who laid the corner stone thereof..."* KJV

In this scripture, it would seem as though God makes reference again to these pillars, but here calls them, "the foundations of the earth;" which, considering the purpose that these pillars served, is an appropriate name for them. God asks Job to explain to Him how these pillars were fastened. That's a good question. Obviously, only God can know that. This was the point God was making, since God is the all powerful Creator, He can do whatever He wants, whenever He wants, and He doesn't need to ask anybody's permission. He is sovereign; He is the *only wise God*.

Additionally, when the fountains of the great deep broke open, and the flood waters began to gush out of the earth, it caused the earth's crust to sink down into the subterranean chamber. This would have crushed and fragmented all of the pillars that supported the earth's crust. Some of these crushed and fragmented pillars were carried out of the subterranean chamber, and launched out of the earth to become asteroids and meteoroids. This would mean that the foundations of the world, which had previously been hidden and concealed, would be "discovered" or revealed. This may be what Psalm 18:15 is referring to:

Psalm 18:15 *"Then the channels of waters were seen, and the <u>foundations of the world were discovered</u> at thy rebuke, O LORD..."* KJV

> ## THESE SUBTERRANEAN WATERS BURST FORTH OUT OF THE EARTH AND BROUGHT ABOUT THE WORLDWIDE FLOOD

Genesis 7:11 *"When Noah was 600 years old, on the seventeenth day of the second month, <u>the underground waters burst forth</u> on the earth, and the rain fell in mighty torrents from the sky." NLT*

In the original Hebrew text of the Bible, the word that we have translated into the English language as "burst forth," is baqa (בָּקַע). This word "baqa" means to violently cleave or rip open. In several places, this word refers to the splitting open of the earth's crust:

Numbers 16:31 *"Then it came about as he finished speaking all these words, that the ground that was under them split [baqa] open." NAS*

Micah 1:4 *"The mountains will melt under Him, and the valleys will split [baqa] like wax before the fire, like waters poured down a steep place." NKJV*

Proverbs 3:20 *"By His knowledge the deep fountains of the earth burst forth [baqa], and the clouds poured down rain." NLT*

It is interesting to note that this word "baqa" has also been translated as "hatch." It describes a baby bird hatching out of its egg in Isaiah 34:15 and Isaiah 59:5. This is an interesting manner in which this word is used, since a baby bird breaks out of its shell as the internal pressure from the bird pressing out against the shell eventually rips the shell open. This is basically the same description of how the earth's crust was ripped open (baqa), by internal pressure from the subterranean water chamber.

The following scriptures also seem to indicate how the fountains of the great deep violently burst open exposing the waters beneath:

Job 38:8 *"Or who shut up the sea with doors when it broke forth and issued out..." AMP*

Psalm 18:15 *"Then the channels of waters were seen, and the foundations of the world were discovered at thy rebuke, O LORD, at the blast of the breath of thy nostrils." KJV*

It is interesting to note that in every instance, the events described in these scriptures are always listed in the same order – that is, first, the fountains of the great deep burst open, and then the rain pours down. It appears to be listed in the same cause-and-effect order as the description of the Hydroplate Theory. Remember, according to the Hydroplate Theory, the source of the rain water was the water from the fountains that burst forth out of the earth, spread throughout the atmosphere, and then fell back down to earth as violent, torrential rain. This cause-and-effect order is positively identified in at least the following two scriptures noted previously – Genesis 7:11, Proverbs 3:20, and also:

Genesis 8:2 *"...the fountains of the deep and the floodgates of the sky were closed, and the rain from the sky was restrained..."* NAS

It is also interesting to note the terminology used for the rain from the flood. The NAS and the NIV translations of the Bible use the term "floodgates" to describe the rain pouring down. The NLT uses the term "torrential rain," and the AMP describes it as "gushing rain." The KJV (and several other versions) simply use the expression, "rain from heaven." However, this word "rain" is translated from the Hebrew word *geshem* (גֶּשֶׁם), which means a *violent down pouring* of rain. The most common Hebrew word that is used in the Old Testament for rain is *matar* (מָטָר). The difference between these two Hebrew words is that *matar* simply means normal rain, where *geshem* means rain that pours down violently. In every instance where the rain from the flood was mentioned, the Hebrew word *geshem* is used. This more aptly explains the reason for the terms such as "floodgates," and "gushing rain," as noted above.

> **GESHEM RAIN POURED UPON THE EARTH FOR 40 DAYS, BUT AFTER THE RAIN CEASED, THE FLOOD WATERS CONTINUED TO RISE UNTIL THE 150TH DAY**

Genesis 7:12 *"And the rain was upon the earth forty days and forty nights."* KJV

Genesis 7:18-19, 24 *"And the waters prevailed, and were increased greatly upon the earth; and the ark went upon the face of the waters. And the waters prevailed exceedingly upon the earth; and all the high hills, that were under the whole heaven, were covered... And the waters prevailed upon the earth* *an hundred and fifty days." KJV*

These scriptures would actually be somewhat difficult to understand without some explanatory theory. If the flood was caused only by water pouring down as rain for 40 days, then how could the waters continue to "greatly increase" and "prevail" over the earth until all of the "high hills... were covered" after 150 days? Obviously, if the water continued to rise without the rain falling, then there had to be another source for the water.

Of course, the Bible does tell us what that source was – the fountains of the great deep. And, the Hydroplate Theory explains this in a way that perfectly harmonizes with the text of the Bible. The explanation is as follows:

1. The fountains of the great deep broke open.
2. This caused geshem rain to pour back down on earth for 40 days and 40 nights.
3. After 40 days and 40 nights, the rain ceased to fall.
4. Water from the subterranean chamber continued to surge out of the earth causing the water level to rise until the 150th day.
5. On the 150th day, the fountains were closed (Genesis 8:2) because the hydroplates settled onto the floor of the subterranean water chamber, choking off, or pinching shut the flow of water.

MOUNTAINS WERE FORMED RAPIDLY ALLOWING THE FLOODWATERS TO RECEDE BACK INTO NEWLY CREATED OCEAN BASINS

Psalm 104:6-9 *"...The waters were standing above the mountains. At Thy rebuke they fled; At the sound of Thy thunder they hurried away. The mountains rose; the valleys sank down to the place which Thou didst establish for them. Thou didst set a boundary that they may not pass over; That they may not return to cover the earth." NAS*

Here, the scripture plainly tells us that the "mountains rose" and the "valleys sank down." This appears to describe the compression event. Another Bible translation says it this way:

Psalm 104:8 *"The mountains came up and the valleys went down..." BBE*

The Hydroplate Theory explains the rapid formation of mountains and valleys during the compression event in the continental drift phase. Can you imagine the shear catastrophic forces that acted upon the earth at this time? It was incredible! What do you think it sounded like when massive hydroplates began skidding and colliding? The thundering crashes were probably deafening. It is possible that this is what the scripture is describing when it speaks of, *"the sound of Thy thunder"* in this passage.

Notice, in verse 9 the scripture indicates that God set a boundary for the flood waters that they could not pass over, so that they could never return again to completely cover the earth. Evidently, these passages of scripture are talking about the global flood, because this is a clear reference to the promise God gave to Noah and the animals, to never again destroy the earth with waters from a flood.

The Hydroplate Theory explains this wonderfully. Since the continents were squished and thickened, high mountains formed rapidly which the water cannot pass over. So, a boundary was set for the water. It does not matter how high the ocean waters rise, they will never be able to overcome the mountains we have today.

Additionally, once the hydroplates settled onto the floor of the subterranean chamber, water was no longer pressed out onto the earth's surface. Then, the earth's surface water would have collected and receded into the large water basins (oceans). Since almost all of the water from the subterranean chamber was released during the flood, and the hydroplates are now almost completely resting on solid foundations, it is completely clear why there will never be another global flood.

BEFORE THE FLOOD, A YEAR PROBABLY HAD 360 DAYS

The Hebrews have historically recognized a 360-day calendar year. Nearly every ancient culture has observed and used a 360-day year, including the Persians, Egyptians, Hindus, Chaldeans, Assyrians, Babylonians, Chinese, Greeks, Romans, Aztecs, Incas, Peruvians, and Mayans.

In fact, the ancient Babylonian astronomers divided a circle into 360 degrees, a custom we still carry today. Have you ever asked yourself, why does a circle have 360 degrees? Other numbers, like 100 or 1000, would have been much easier to use. Apparently, ancient people chose 360 degrees to match the 360 days of the year, which was familiar before the time of the flood.

The Bible plainly indicates that a preflood year consisted of 360 days.

In Genesis 7:11, the Bible states that the flood began on the 17th day of the 2nd month. Then, in Genesis 8:4, it states that the ark came to rest on the top of the mountains of Ararat on the 17th day of the 7th month. These two dates are exactly five months apart. In Genesis 7:24, the Bible tells us that the floodwaters prevailed over the high hills for 150 days. Naturally, if the ark was resting on a mountaintop, then the waters were no longer "prevailing" by this time. So, the Bible gives us two expressions for the length of time of the flood: 5 months and 150 days. If we divide 150 days by 5 months, we get a month with exactly 30 days. If we multiply a 30 day month times 12 months, we get a year with exactly 360 days. All of this is within the text of the Bible.

Additionally, these timeframes seem to be indicated within prophetic verses of scripture. The "time, times, and a half of time," as mentioned in Daniel 7:25, Daniel 12:7, and Revelation 12:14, where a "time" is the equivalent of a *year*, would perfectly equal the same period as the 42 months (Revelation 11:2) and 1,260 days of the Revelation (Revelation 11:3, 12:6), if a year consisted of 360 days.

If the Bible seems to indicate an original 360-day year, then why is our current year 365 ¼ days long? In order for our year to change, earth's spin rate would have to increase. Most likely, the events of the flood, as explained by the Hydroplate Theory caused this increase in the earth's spin rate, which probably caused a change in our year from 360 to 365 ¼ days.

For another picture of how the Hydroplate Theory harmonizes with the text of the Bible, see the table on the following page (which was taken from *In the Beginning*).

Comparison of Biblical Chronology
with the Major Events of the Hydroplate Theory

Biblical Chronology	Hydroplate Theory
Day 2 of Creation Week: The earth was covered by water (Ge. 1:2). Then a "firmament" was created which separated the liquid water above from the liquid water below.	The initial condition is established: a layer of water is formed underneath the earth's crust–the firmament.
The waters below the heavens are gathered into one place, and then the dry land appears (Ge. 1:9).	A rock crust, resting on a layer of water, will automatically deform. Portions of this crust subsided down to the floor of the subterranean chamber forming "pillars," while other portions bent upward. Water above the crust, drained into the depressions or basins that were formed on the crust, and the dry land appeared.
The flood begins suddenly when all of the fountains of the great deep break open on one day (Ge. 7:11). Geshem rain begins to pour.	**Rupture Phase:** A crack propagates around the earth in 2-3 hours, releasing subterranean water. Some fountains of muddy water jet high above the earth. Mammoths are frozen in muddy hail falling from above the atmosphere. The highest velocity water escapes earth and forms comets. Launched rocks become asteroids and meteoroids.
40 days and 40 nights of geshem rain ends (Ge. 7:4,12) Flood waters rose until the 150th day, when they covered all preflood mountains (Ge. 7:19-24)	**Flood Phase:** Rising flood waters blanket and suppress the high jetting waters of the fountains of the great deep. Animals and plants are buried in sediments from the muddy water. High pressure water continues to gush up into the flood waters. A known physical process called liquefaction sorts sediments and dead plants and animals. Coal and oil deposits form.
150th day: God causes a wind to pass over the earth and the waters begin to subside. The ark comes to rest on the mountains of Ararat (Ge. 8:1-4).	**Continental Drift Phase:** Mid-Atlantic Ridge buckles upward; Atlantic floor rises and the Pacific region subsides, so that the hydroplates begin to slide downhill, sliding on a layer of lubricating water. When the massive hydroplates decelerate, they crush, thicken, buckle, and heat up in a gigantic compression event. Continents take on their present shape. As major mountains form, air is displaced causing a great wind. The earth slowly rolls 35º-45º and the poles shift.
150th-371st day: Noah and his family remain in the ark. 371st day: the ark is off-loaded (Ge. 8:15-19) 371st day until the present time	**Recovery Phase:** Hostile environment: earthquakes, inner earth heated, oceanic trenches form, volcanoes erupt, water drains, high continents settle, vegetation is reestablished and Ice Age begins. Lowered sea level allows people and animals to migrate all over the earth. Plateaus form and large canyons begin to form when water breeches natural dams.

A Concise Overview

The purpose of this short book is to give the reader a brief explanation and introduction to the Hydroplate Theory. For those readers who are still very much intrigued by this topic, I would highly recommend that you read *In the Beginning* by Dr. Walt Brown. All of the information in this book was derived and excerpted from *In the Beginning*. There is still so much for you to learn. In fact, many mysteries were purposely left unexplained in this little booklet to pique your interest in hopes that it will drive your thirst for more knowledge.

When you read *In the Beginning*, you will learn about exact scientific reasons and physical laws that dictated the events of the four phases of the Hydroplate Theory. Additionally, you will find the explanations to over 25 mysteries that we did not discuss in this book including:

- Earthquakes
- Magnetic variations on the ocean floor
- Ice Age
- Overthrusts
- Limestone
- Salt Domes
- much more

Furthermore, you will find expanded chapters that each give great and detailed information and thoroughly explain how the events of the Hydroplate Theory solve the following mysteries:

- The Origin of Ocean Trenches
- Liquefaction: the Origin of Strata and Layered Fossils
- The Origin of Limestone
- Frozen Mammoths
- The Origin of Comets
- The Origin of Asteroids and Meteoroids

- The Origin of the Grand Canyon

Each of these chapters explains in great detail how the flood would have caused these mysteries, and they also describe the great errors and faults in widely-held evolutionary-based theories which attempt to explain these same mysteries.

The book contains over 30 scientific predictions that have been made based on the principles of the Hydroplate Theory, and they are plainly explained and recorded.   Some of them have already come to pass. Additionally, Part 1 of the book makes a very thorough, *Scientific Case for Creation*, and Part 3 of the book answers the following frequently asked questions (you may have many of these questions yourself):

- How Can the Study of Creation Be Scientific?
- Have New Scientific and Mathematical Tools Detected Adam and Eve?
- Because Galaxies Are Billions of Light-Years Away, Isn't the Universe Billions of Years Old?
- Why Does the Universe Seem To Be Expanding?
- If the Sun and Stars Were Created on Day Four, What Was the Light of Day One?
- How Old Do Evolutionists Say the Universe Is?
- What Was Archaeopteryx?
- How Accurate Is Radiocarbon Dating?
- How Could Saltwater and Freshwater Fish Survive the Flood?
- What about the Dinosaurs?
- Have Planets Been Discovered Outside the Solar System?
- Did the Flood Last 40 Days and 40 Nights?
- Is the Hydroplate Theory Consistent with the Bible?
- How Was the Earth Divided in Peleg's Day?
- Did It Rain Before the Flood?
- What Triggered the Flood?
- If God Made Everything, Who Made God?
- Did a Water Canopy Surround Earth and Contribute to the Flood?
- How Did Human "Races" Develop?
- According to the Bible, When Was Adam Created?
- Is There Life in Outer Space?
- Is There a Large Gap of Time between Genesis 1:1 and 1:2?
- Is Evolution Compatible with the Bible?
- Does the New Testament Support Genesis 1–11?
- How Can Origins Be Taught in High School or College?

- What Are the Social Consequences of Belief in Evolution?
- How Can I Become Involved in This Issue?
- How Do Evolutionists Respond to What You Say?
- How Do You Respond to Common Claims of Evolutionists?
- Why Don't Creationists Publish in Leading Science Journals?

If there really was a worldwide flood, what would we expect to see? The first thing we would expect to see are remnants of living organisms that are now dead and buried. Of course, we have found literally billions of such dead creatures (fossils) buried in rock layers all over the earth.

We would also expect to find evidence of this disaster all over the world which can easily be explained as the result of such a cataclysmic flood. In review, let us consider the following things that the Hydroplate Theory describes as being the result of this flood:

1. **The Mid-Oceanic Ridge** was formed as portions of the chamber floor, no longer pressed down by the overlying crust, buckled upward.
2. **The continental shelves and slopes** were formed as the jetting fountains eroded the sides of the hydroplates all over the world.
3. **Ocean trenches** were dramatically formed when the material from the inside of the earth shifted toward the Atlantic to compensate for the rising Atlantic floor. This shifting pulled parts of the western Pacific down forming the trenches.
4. **Submarine canyons** were formed when water draining off the continents washed down channels in the continental slopes.
5. **Major mountain ranges** were formed in hours when massive hydroplates ran into resistances and began to skid, compress, and buckle. These hydroplates literally smashed themselves as they ran into resistances and thickened, and in some cases folded over on themselves, creating the patterns we see in the sides of mountains today. Additionally, many of the erosion marks we see on mountain chains today came from the drainage of the flood waters.
6. **Volcanic activity began and magma was formed** by frictional heating of sliding rock within the earth.
7. **The earth rolled** by about 35-45 degrees when rapid formation of mountains threw off the earth's spin balance.
8. **Comets and asteroids were launched** into outer space when rocks, debris, and water exploded violently and supersonically out of the earth. Some of the major impact features on the near side of the moon are the result of impacts from debris that launched out of the earth.

9. **Erosion channels on Mars** were formed when comets from the earth crashed onto the Martian surface creating brief saltwater flows that eroded these "channels."

10. **Plateaus were formed** as mountains gradually settled into the subterranean chamber causing the adjacent land to rise.

11. **The length of the year changed** from 360 days to 365 ¼ days due to the changing of the earth's spin rate as a result of the events of the flood.

These are just a few of the mind-numbing catastrophic results of the fountains of the great deep bursting forth in the days of Noah. Dr. Brown has calculated that the energy release of the fountains of the great deep breaking open exceeded the energy from the explosion of 10 billion hydrogen bombs.

As I have said before, so will I say again, "God was angry!" It is very disheartening to see, in this generation that we live in, the desire for people to make God nothing more than a benevolent provider of our wishes and desires. People want God to be nothing more than loving and merciful. Everybody wants to tell us that "God is love, God is love, God is love." While this is certainly true, and while there is no one more loving, kind, and merciful than God, it is also true that He is a God of judgment! God is just and God is holy, and sometimes His righteousness and His justice demands that He bring judgment to a world full of sinners. He did it once before, and the Bible tells us that He will do it again – but this time, with fire.

The only hope that we have, and the only hope that we need, is the opportunity that has been afforded us to be forgiven of our sins by the blood of Jesus Christ! It is only by God's forgiveness and mercy through Jesus Christ, that anybody can escape the coming judgment!

Appendix:
Creation Resources

Thankfully, we are living in a time when there are hundreds of creation based resources. Those mentioned below only make up a small portion of the great amount of material available. Take your time, and learn as much as you can–it will be worth it.

BOOKS ON CREATION

In the Beginning, **Dr. Walt Brown**
This book is probably the best single source for creation material as a whole. It is an absolute must-read!

The Genesis Flood, **John C. Whitcomb and Henry M. Morris**
This book is credited as being the catalyst that ignited the modern day young-earth creation movement. Some portions of the book can be quite technical for the average reader.

Scientific Creationism, **Henry M. Morris**
The editor and main contributor to this book is Henry Morris, who is the founder of the *Institute for Creation Research.* He has written over 50 books (primarily about Creation and science) and authored hundreds of articles.

The Lie: Evolution, **Ken Ham**
Ken Ham is the founder of *Answers in Genesis,* the home of the large *Creation Museum* located near Cincinnati, OH. This book does an excellent job of pointing out the real purpose of evolution.

The Answers Book, **Various Authors**
This book is compiled by various authors from the *Answers in Genesis* staff, and it answers many of the questions that people often ask, such as "where did Cain get his wife," and "how did all the animals fit on the ark?"

Unlocking the Mysteries of Creation, **Dennis R. Petersen**
This is a very nice, full-color book that is great to use for homeschoolers as a supplemental science book. The bright and colorful graphics will hold your attention and keep you turning the pages.

What is Creation Science? **Henry Morris and Gary Parker**
This readable book is an excellent resource for those who need to reexamine the subject of origins, especially if they wonder, "what difference does it make" or feel that "science favors evolution."

CREATION BASED ORGANIZATIONS

FOUNDATION FOR CREATION DOCTRINE
PO Box 13008
Fresno, CA 93794
(559) 276-9777
fax (559) 274-9777
FCD was founded by Diego Rodriguez, the author of this book. This ministry is primarily focused on distributing resources, producing new material, and teaching seminars.

CENTER FOR SCIENTIFIC CREATION
5612 N. 20th Place
Phoenix, AZ 85016
(602) 955-7663
www.creationscience.com
CSC is the home of Dr. Walt Brown and the Hydroplate Theory. The newest, updated version of his book is always available online, as well as a short animation of the theory.

INSTITUTE FOR CREATION RESEARCH

10946 Woodside Ave.
Santee, CA 92071
(619) 448-0900
www.icr.og
ICR was founded by Henry Morris, one of the co-authors of *The Genesis Flood*. ICR staff members have written dozens of books, and have performed many creation-based research projects. They also have a small creation museum in San Diego, CA.

ANSWERS IN GENESIS

PO Box 510
Hebron, KY 41048
(800) 778-3390
www.answersingenesis.com
AiG was founded by Ken Ham, author of several books about creation. AiG is also the home of the *Creation Museum*, and the source for a quarterly creation magazine called *Answers*.

CREATION RESEARCH SOCIETY

Creation Research Society
P.O. Box 8263
St. Joseph, MO 64508-8263
(877) 277-2665
www.creationresearch.org
CRS publishes several books and also a quarterly journal. Many resources are available through this ministry.

CREATION SCIENCE EVANGELISM

29 Cummings Road
Pensacola, Florida USA, 32503
(877) 479-3466
www.drdino.com
CSE was founded by Kent Hovind, also known as Dr. Dino. Dr. Hovind is most well known for his seminars which he teaches all over the country, as well as his many debates. Many creative resources, such as fossil replicas, are available from CSE.

Illustration Credits

Most of the pictures used in this book can be obtained in electronic format or as overhead transparencies by contacting the **Center for Scientific Creation** (contact information is on page 74).

Figure 1:   Cropped from figure 2, page 33

Figure 2:   World Ocean Floor, Bruce C. Heezen and Marie Tharp, 1977, copyright Marie Tharp 1977, reproduced by permission of Marie Tharp Oceanic Cartographer, 1 Washington Ave., South Nyack, NY 10960, page 34

Figure 3:   Bradley W. Anderson, page 35

Figure 4:   Digital Wisdom, Incorporated, (labels by Diego Rodriguez) page 36

Figure 5:   Reproduced with the permission of the Minister of Public Works and Government Services Canada, 2006 and Courtesy of Natural Resources Canada, Geological Survey of Canada (photo no. GSC180345), page 38

Figure 6:   Illustration by Allen Beechel, page 39

Figure 7:   Bradley W. Anderson, page 40

Figure 8:   Bradley W. Anderson, page 41

Figure 9:   Drawing by Steve Daniels, page 42

Figure 10:   Bradley W. Anderson, page 44

Figure 11:   Steve Daniels, page 45

Figure 12:   Steve Daniels, page 46

Figure 13:   Bradley W. Anderson, page 47

Figure 14:   Bradley W. Anderson, page 48

Figure 15:   Bradley W. Anderson, page 49

Figure 16:   Bradley W. Anderson, page 50

Figure 17:   Bradley W. Anderson/Steve Daniels, page 51

Figure 18:   Computer generated map by Bradley W. Anderson, page 56

ORDER FORM

If you would like to order more copies of this book,
please feel free to contact us, or send in a copy of this form.

Qty	Title	Price	Total
___	The Fountains of the Great Deep .	$8.95	_____
___	In the Beginning	$24.95	_____
___	In the Beginning CD . . . . (contains PDF version and short animation)	$12.95	_____
		Subtotal	_____
		Shipping & Handling (15%)	_____
		TOTAL	_____

You may send us this form by fax or mail.

❏ Check Enclosed (make checks payable to Sound Alive Publishing)

❏ Credit Card (statement will show charge from Faith Tabernacle Bookstore)

Name on Card _____

Credit Card # _____

Expiration Date _____

Signature _____

PO Box 13008, Fresno, CA 93794 • fax (559) 274-9777

About the Author

Diego Rodriguez is Pastor of Faith Tabernacle church in Fresno, CA. He is the founder of the *Foundation for Creation Doctrine*, a ministry dedicated to promoting and teaching the truth of a literal, 6-day creation. This message is taught through books, seminars, and other related resources distributed by the ministry.

Diego shares the joy of the ministry with his wife, LaReina, and their five homeschooled children: Moriah, Miranda, Marissa, Micaiah, and Manoah.

Foundation for
CREATION
D O C T R I N E